NER
press

Farewell Fear

Farewell Fear

Theodore Dalrymple

Published by New English Review Press
a subsidiary of World Encounter Institute
PO Box 158397
Nashville, Tennessee 37215

Cover Design by Kendra Adams

ISBN: 978-0-9854394-7-7

First Edition

 NEW ENGLISH REVIEW PRESS
newenglishreview.org

Contents

So farewell hope, and with hope farewell fear, Farewell remorse!
All good to me is lost; Evil, be thou my good.

- Milton, *Paradise Lost*

Preface

R eflection on the small change of life soon leads to deeper questions – at least, such has long been my belief, or perhaps my hope. It is also my hope that the following short essays, published by the *New English Review* over a period of three years, bear out my belief.

We take a lot of pride, partly justified, in the fact that ours is the age of information. It is certainly true, as I have found, that it is now possible to do in an afternoon, comfortably sitting in one's room, research that would once have taken months, if it could have been done at all. For example, I no longer have to trudge to the library to look up a single entry in the *Dictionary of National Biography*: what once took up a whole afternoon, and sapped my energy, both prospectively and retrospectively, now takes considerably less than five minutes. It is wonderful.

But information without perspective – here I shamefacedly admit that I am quoting myself – is a higher form of ignorance. Information, be it ever so copious, will not by itself result in truth, much less in wisdom. Reflection on the meaning of information is at least as important as the information itself. The best informed man is not necessarily the wisest, therefore, or the surest guide to any subject.

I hope these little essays conduce to wisdom and intellectual honesty (another frequent casualty of a surfeit of information).

I am grateful to Rebecca Bynum on a number of grounds: first for having asked me to write for the *New English Review*; second for suggesting that my essays might be put in a book; and third for her editorial assistance. Thank you.

1
What the Hedgehog Knows

B eing a scholar of nothing, I allow my intellectual interest to wander hither and yon. Or perhaps it is because I allow my intellectual interest to wander hither and yon that I am a scholar of nothing. Be that as it may, I admire specialists and am grateful to them for their researches, but I could never be one myself. Whenever it is imperatively necessary for me to read a book pursuant to something that I am currently writing about, I immediately lose interest in it, as I lose my appetite in a restaurant if I wait too long for the food to arrive; and then I want to read about something else entirely.

And so it was when, at a time when I was supposed to be writing about the methadone treatment of heroin addiction (methadone kills more people in Dublin than heroin), I wandered into a bookshop in Lower Baggot Street in Dublin that sold cheap remaindered books, that I bought a book about hedgehogs. It wasn't a very good book, as it turned out, and was written in that jocular fashion in which people who are enthusiastic or passionate about something nowadays feel it necessary to write, for fear of appearing solemn or of revealing too much of themselves. But the book was just what I wanted to divert me from what I ought to have been reading.

I have always found hedgehogs rather appealing, though I am afraid that (before I bought this book) my knowledge of them had not advanced very much beyond that acquired when I read *The Tale of Mrs*

Tiggy-Winkle at the age of six. Of course, I knew that hedgehogs did not really do the laundry for other creatures such as wrens and rabbits; and from personal observation I knew that, far from being clean, they were often alive with fleas. But I did not know that those fleas were specific to hedgehogs; numbered on average 100 per host; appeared so prominent to casual observers such as I because the hedgehogs that most people encounter are ill and the fleas are preparing to abandon the sinking insectivore, as it were; or that the flourishing hedgehogs of New Zealand (that were introduced into the country in the nineteenth century to Anglicise its landscape and fauna) do not have fleas, and that therefore the fleas serve no symbiotic purpose for the hedgehog.

There were other facts I learned, for example that the top speed of a hedgehog is five and a half miles an hour, though the book did not inform the reader how long it can keep it up. Even in my early sixties, my top speed is probably quite impressive; less so is the number of yards for which I can sustain it. I learnt also that the hedgehog wanders up to a mile and a half at night, and that it hibernates not from metabolic necessity to avoid the cold but because of the decline in its food supply.

Part of the delight of these facts for me was their complete irrelevance to anything else with which I normally concern myself. This reassured me that my interest in the world is, at least in part, disinterested, that I like knowledge for its own sake: and since knowledge for its own sake is a noble thing, that I am in part, or at least sometimes capable of being in part, noble.

But the book also promised (though it did not really deliver) an insight into the peculiar human world of hedgehog enthusiasts, of whose existence I became aware quite by chance.

A friend of mine, aware that I did not have a television, desired to prove to me the genius of the British comedian, Sacha Baron Cohen, and therefore sat me down to watch an episode of the television series in which Baron Cohen plays Ali G, a suburban white boy who wants to be a black ghetto boy. And Baron Cohen mastered the gestures and patois of the ghetto culture brilliantly.

Many prominent people were invited to interview by Ali G; the circumstances and reasons for the interviews were grossly misrepresented to them, and in the event they were often made to look very foolish since they did not understand the joke, or indeed that it was a joke. Some people might say that public figures such as politicians were fair game, that they could have been expected to look after themselves; but if we expect politicians not to deceive us, surely it is incumbent on us not

to deceive them. The joke was a good one (if quickly grasped and therefore limited); but the extraction of a laugh at someone's expense, even if
that person is not estimable, procured by unethical behaviour, does not
justify the unethical behaviour necessary to extract it. Entertainment is
not the only good, much less the highest good.

In any case, the people satirised by Ali G in the episode which
my friend showed me were not public figures; they had not sought the
limelight, nor were they the kind of people who in any way deserved
cruel mockery. They were obviously kind and harmless, completely naïve in the ways of television or show business. They were people who
went round the countryside rescuing injured hedgehogs, restoring them
to health and returning them to nature. (One of the questions that the
book answered was whether rescued hedgehogs survive when returned
to nature, the answer being in the affirmative).

The kindness and innocence of the hedgehog rescuers was obvious and yet they were held up to ridicule, only that millions of people
might have a moment's laughter. I thought this repellently exploitative
and unscrupulous: satire is often necessarily hurtful to the satirised, but
its object should be worthy of satire. There is not so much kindness and
innocence in the world that it should be publicly mocked.

In fact, it is rather difficult to believe that those who interest themselves practically in the welfare of hedgehogs can be bad people. When
I bought the book, the man at the till – who had a kind and gentle
face himself – said that he had thought the book might be interesting.
I mentioned that there were surprisingly many people who interested
themselves in hedgehogs, and that they were probably very different in
character from those (of whom there were also many) who interested
themselves in snakes.

'I would imagine so,' said the bookseller. 'Personally, I would be
more inclined to the hedgehog people than to the snake people.'

Indeed so; though the generalisation behind his preference would
be open to exception. I have known in my life only one professional her-
petologist, and he was a splendid, if somewhat reckless, character. And
recently I read the autobiography (called *A Venemous Life*) of an Austra-
lian doctor, Professor Struan Sutherland, who alas died comparatively
young of a rare neurological condition, who developed anti-venoms to
the poisonous creatures of Australia. Sutherland was a fine man, humorous, self-deprecating and dedicated to saving people's lives. No one has
died of the bite of the funnel-web spider – said to be the world's most
dangerous spider - since he developed the anti-venom to it.

But nevertheless, the generalisation probably holds: a visit to a pet shop specialising in snakes and gila monsters, and observation of its clientele, will be enough to convince the average person of this.

No doubt there will be many people (other than Ali G) who find the notion of hedgehog rescue ridiculous. They will say, rightly, that there are many more important problems in the world than that of injured hedgehogs, or even of the decline in hedgehog numbers. And, of course, they are quite right. When one reads that more than a million people a year die of malaria, the fate of the hedgehog seems not of the first importance.

But should people devote themselves only to the things that are of the most importance? What would a society be like in which people did so? It would not be much fun, I imagine.

Let us, for the sake of argument, say that we could rank all the problems in the world in a rational order of importance, from most to least. This, of course, would require what is almost certainly impossible to find, a common measure of importance: a measure impossible to find for a number of philosophical reasons, among them that importance is a non-natural quality ascribed to facts about the world by a mind that finds meaning in them according to its own lights. But, I repeat, let us just suppose that there were a Richter scale of the importance of problems.

The rationalist would say that human energy should be directed at those problems in proportion to their importance. If we decided, for example, that the problem of malaria was 100,000,000 times more serious than that of the decline of the numbers of hedgehogs in the wild, then it would follow that 100,000,000 times more energy should go into the solution of the former than into the solution of the latter.

If this were the case, it is probable that no effort whatsoever would go into the preservation of hedgehogs, which might then die out. And I feel, though I cannot prove, that a world without hedgehogs would be a slightly impoverished world. I acknowledge that there is an element of inconsistent nature mysticism in this – I certainly do not feel the same about the possible extinction of all other species, of *Ascaris lumbricoides*, for example, the large roundworm that inhabits the intestines of humans, particularly children, and that not only causes ill-health but is aesthetically disgusting – but if we must justify our feelings from unshakeable and utterly consistent first principles we shall soon have no feelings left to justify. And, of course, once the hedgehog is extinct it is – barring advances in technology that are for the moment in the realm of science fiction – extinct forever. No one will ever again have the pleasure of seeing a live dodo.

Now the fact is that a good and enjoyable human life is composed of many small pleasures. A man who devotes himself to raising herbs for cooking, for example, is not engaged upon the most important work in the world, and probably wouldn't even claim to be, though it might occupy most of his energy or thoughts; but the world would be a poorer place without his herbs. The same is true of the vast majority of human activities.

Just as it is beyond human or any other known capability to organise a command economy, such that goods are produced according to rationally preconceived need and not according to any other criterion, so it would be impossible to organise a society in which people allocated their energies according to the importance of activities that were likewise rationally preconceived. Such a society would be Aspergerish in its disregard of human qualities and desires; it would exclude thousands of worthwhile, but individually unimportant, activities. Stamp collecting for example, seems to many the acme of futility, notwithstanding the great erudition of stamp collectors on their subject: but would the world really be a better, richer place without it, or if the stamp collectors were forced to turn their attention to something deemed more important as judged by rational criteria? (And who, exactly, is to do the deeming? A rationalist is a person who proclaims his own preferences to be metaphysically sound.)

There are those who would go beyond ordering human energies within societies to ordering human energies between societies. For example, they criticise the fact that medical research is not organised and funded according to the importance of the subject researched, judged by world-wide criteria. If disease x causes y per cent of world mortality, then y per cent of funds for medical research should be spent on research into disease x. No considerations of commercial viability, practical feasibility or scientific or intellectual importance, to say nothing of avowals of ignorance, should be permitted to enter the allocation of resources.

The fox, says the proverb, knows many things, but the hedgehog knows one big thing. What is the one big thing that the hedgehog knows? It is that there is not only one big thing to know.

2
When Irish Eyes Are Crying

By far the most revealing comment on the Irish crisis was made - inadvertently, of course - by the Taoiseach (Prime Minister) of Ireland when he announced that he would resign, dissolve parliament and hold a general election in the near future, once the austerity budget that was the condition of a financial bail-out of his country had been passed.

> 'There are occasions,' he said, 'when the imperative of serving the national interest transcends other concerns, including party political and personal concerns.'

Well, that's nice to know: there are occasions when the needs of the country may be permitted to interfere (though not, of course, for very long) with a politician's career plans. But think how galling it must be for him, poor fellow, when this happens! There will soon be a name for the psychiatric condition such occasions cause in politicians: Politician's Self-Sacrificial Stress Disorder.

It is too late for this condition to get into the fifth edition of the Diagnostic and Statistical Manual of the American Psychiatric Association, of course, but there will surely be room for it in the sixth, which will probably be as long as, or longer than, the tax regulations. Let us, in preparation for the publication of that great compendium, develop a few diagnostic criteria for PSSD.

First, as a *sine qua non*, the sufferer must be a full-time professional politician, must have been so for at least four fifths of his adult life, and must never have worked in any productive capacity whatever. PSSD is therefore an occupational psychiatric disorder.

Second, the sufferer must have been in high office when a political or economic crisis struck the country, that anyone with an IQ higher than 80 would recognise as having been such, and that was of far larger scale or more grave than any crisis in the last fifty years. (What about forty years, you ask? No, the scientific committee charged with enumerating the criteria has laid down fifty, so fifty it must be.)

Third, the sufferer must have developed at least two of the following four symptoms within sixth months either of the unmistakable presence of the crisis, or of resigning office, whichever was the later (usually the latter, of course):

a)   A preoccupying self-pity
b)   An ability to rationalise everything he did while in power
c)   A tendency to blame his successors for the subsequent mess in the country
d)   A tendency to blame his predecessors for the subsequent mess in the country.

The above symptoms must be deemed to be in excess of the normal human tendency to display them, and should where possible be measured objectively by validated instruments, for example Williams and Greenbaum's Self-Pity Checklist Questionnaire (498th Edition), which has a cut-off point of 37/48. (It also contains questions, such as 'Do you blame your parents for everything?' which measure the truthfulness of the respondents. Those who answer 'No' to the question 'Do you blame your parents for everything?' are lying.)

In addition to the above, the sufferer must display at least one the following symptoms, though most typically he will display both:

a)   A tendency to continue to interfere in public affairs despite the self-evidently catastrophic nature of his previous efforts in this direction
b)   A large publisher's advance for his 700 page memoirs, which will remain unsold in the bookshops, or unread if transported therefrom into private houses.

However, it is time now to turn our attention from the sunny uplands of scientific psychology, where luckily the air is so clear that everything can be seen and measured, to the murky swamps of politics and economics.

We read in the press (and I know from having travelled there recently) that the crisis is felt in Ireland to be more than a merely economic one, but existential too. This is because, while everyone recognises that the Irish political class behaved abominably, as did the bankers, the fact is that, with few exceptions, the population did not exactly emerge covered in glory either. After all, you can lead a man to a loan, but you can't make him borrow; and the fact is that the Irish population behaved as if they believed that house prices had escaped the pull of the earth's gravity. Not to put too fine a point upon it, the Irish were fools.

But if the Irish have nothing to congratulate themselves about, neither should they consider themselves worse in this respect than many other people I could name. They, too, these other people, had the luck of the Irish, if luck is quite the word I seek.

One thing I have noticed of late is that the sums mentioned for the Irish debt vary considerably, both within and between newspapers, and sometimes even in the same edition of the same newspaper. But I suppose when the sums are so enormous, it doesn't really matter what they are down to the last ten billion. For example, in the *Financial Times* of November 22, 2010 I found that British banks had lent $136 billion to Ireland, but by the following day it had risen to $149 billion in the same newspaper, without (so far as I know) any extra loans having been made.

According to one article in the *Financial Times*, Ireland's gross external debt was $2,131,000,000,000 (if I have added the right number of noughts, one is never quite sure these days). Not bad, for a population of just over 4 million to run up in a handful of years: which is to say, if my mental arithmetic serves me right, very nearly $500,000 per head of population. Of course, it is possible that I am out by an order of magnitude or two, but at these kind of figures that doesn't seem to matter all that much either. It makes me rather ashamed of my own paltry finances.

Anyway, here are the figures for bank lending to Ireland, according to the *Financial Times* for 23 November: United Kingdom $149,000,000,000, Germany $139,000,000,000, United States $69,000,000,000, Belgium $54,000,000,000, France $50,000,000,000, Japan $27,000,000,000 and others $229,000,000,000, for a grand total - not grand in the Irish sense of 'That's excellent,' incidentally - of

$731,000,000,000.

This amounts to bank loans of about $175,000 for every man, woman and child in the Republic. It is clear that it is not the Irish alone who have been foolish, therefore. It takes two (at least) to make a bad loan. There can hardly be a major bank in the whole world, save the Canadian, that has behaved prudently, though there must be, and we know for a fact that there were, many that must have behave imprudently, not to say recklessly.

What could be said in defence of the banks? I can't really think of very much. When they lend money, which is not after all theirs, they have a duty to do so prudently. I have very little idea of how bankers spend their day, but I had always rather assumed that at least part of their work must be checking whether those who want to borrow money from their banks have any realistic chance of re-payment, and how good the collateral is that they offer. Perhaps the banks didn't bother with this once they became too big to fail; but I rather doubt (though I might be wrong) that anyone sat round and said, 'Don't worry, the taxpayer will always foot the bill if default threatens us.'

It might be said, I suppose, than no one had the overall picture as they have now, and that therefore the banks acted in ignorance and good faith. But this doesn't really wash, because one bank, the Royal Bank of Scotland, lent by itself $50,000,000,000, that is to say $12,000 for every man, woman and child in the Republic. There was not the faintest hint of caution in its exposure, though I suspect that if you asked the man in his street, even one who was himself deeply indebted, whether he thought it is a good idea for a bank to lend more than $12,000 to every man, woman and child in a country, he would have said, without needing to know more, that it was not.

The shallowness of some of the commentary on Ireland by economists has startled me. This commentary has repeated many times that the Irish problem is private indebtedness, not appalling public finances as in Greece. This is said because, until very recently, the Irish budget was more or less balanced.

What this fails to notice is that the bubble created by the private indebtedness was used to inflate both the size and remuneration of the public service in Ireland, no doubt partly to create a grateful political clientele. The appalling nature of the Irish public finances was merely postponed, however, until the whole Ponzi scheme collapsed. Government expenditure rose by more than 50 per cent in a few years; the tax revenues to finance this came largely from property sales taxes that also entered a bubble, along with the property market. Now that the bubble has collapsed, the Irish state has been left with enormous wage and pension obligations which it cannot possibly meet from current or likely future revenues. That is why both private and public demand have entered a downward spiral.

Britain is in the same position, though the fact that the economy is so much larger and more complex has disguised it somewhat hitherto. Moreover, having kept its own Monopoly-money currency, the pound, it can inflate itself out of debt if it needs to, and probably will. Inflation has harmful social and psychological consequences, but it is at least an option. Unfortunately for Ireland it has a currency whose value it cannot control.

The Irish are right to think that the crisis is more than economic, that it is existential; but they are wrong to think that it is specifically or mainly Irish. It is, in fact, a crisis of western man who cannot control his appetites, who wants today what only the labour of the future can supply, or supply honestly. Western man is, in effect, a child.

In my own country, for example, there has been a decisive shift in attitude to debt within my lifetime. The British people are now among the most incontinent and childish in the management of their own affairs of any people in the world, which is why they are so deeply indebted for what are, essentially, trifles. It is within my memory that people took pride in not buying what they could not afford; they feared debt as if it were a disease. The idea of repudiating debts, of simply walking away from them, was totally alien to them. They would not behave in this fashion, not because it was illegal, but because their self-respect would not allow them to do so. How can you hold your head up after you had indulged in an elaborate form of theft?

What had effected this change? I suspect that the decline of religion, both as a system of belief and a system of social control, has something to do with it. (Is it really a coincidence that the Irish crisis has struck at precisely the same time as the total evaporation of Catholicism's influence in Ireland?) I say this as someone without religious belief. But where there is no belief that life has transcendent purpose, that there is in effect more to life than this life itself, it is hardly surprising that people - that is to say, many people - take as their philosophy '*Après nous, le déluge.*' The problem is that the deluge may not be *après nous*.

Germany is an interesting exception to the *après nous* philosophy. I remember that it isn't so long ago that I read articles in respected publications pitying the poor old Germans for being so unimaginative as still to make things: manufacturing was the wave of the past, while taking in each other's washing was the wave of the future, where all the easy money was to be made.

3
Mumbai's the Word

I once stayed at the Taj Mahal in Bombay (as it was then still called). I didn't enjoy it as much as I might have done, because I was recovering from the hepatitis I had contracted in the South Seas. But I still recognised the magnificence of the institution, even as I regretted the modern excrescence that been added to the original building and that ruined its architectural unity.

One morning as I left the hotel, a middle-aged man with a black umbrella and a medium-clean *dhoti* said to me, 'Come with me.' His teeth were stained with betel, but I thought to myself, 'Why not?' and so I went with him. I don't know whether he recognised in me a man with a taste for the unusual and the bizarre, but if so he was a man of sound judgement. As for me, I guessed that he was odd rather than bad, and I proved to be right as well.

Before long, we were winding our way through narrow alleys (I cannot think of a better word for them) through slums in which the shacks were made of every cast-off material, and I must have seemed like an apparition to the inhabitants. The mud beneath my feet was not merely mixture of earth with water, but with every fluid known to man.

Our first port of call was a special crematorium, where Hindu rites were carried out. It was only for legs: and not just any legs, but those that were amputated by trains when the people riding on their roofs (as untold thousands, perhaps millions, did, every day) fell off and a passing train severed their lower limb or limbs. The crematorium never lacked for activity - I hardly like to call it work or business.

Was my guide - for so I now considered him - merely ghoulish, or was he trying by subtle indirection to convey to me the desperation of the life of the Indian poor? (Of course, the surfers of the trains were not the poorest of the poor: they had definite work to go to.) Was he, perhaps, a reader of Emily Dickinson?

Tell all the Truth but tell it slant -
Success in Circuit lies
Too bright for our infirm Delight
The Truth's superb surprise
As Lightning to the Children eased
With explanation kind
The Truth must dazzle gradually
Or every man be blind -

The truth to be conveyed here was not superb, perhaps, but it was, in certain sense, dazzling - or would have been had I never visited India before.

We did not linger long over the legs, however, but progressed on to the babies. As we continued to thread our way through the noisome alleys my informant told me, whether or not accurately from the purely doctrinal point of view I was in no position to tell, that those that died before a certain age were not cremated but buried.

Doctrinally correct or not, we soon reached a small piece of ground in which babies were buried. My guide began to poke about in the soil with the tip of his umbrella, and soon came across the delicate skull of an infant.

'You want for souvenir?' he asked, with a total absence of sentimentality about human remains.

I declined, and our tour was over. Having completed my education (if that is what it was), he guided me back to the Taj and left me at the entrance. He did not ask for money but I gave him some that immediately, in a kind of balletically smooth movement, disappeared into a fold in his *dhoti*. I think that if I had not given him any he might not have protested, but simply put it down to fate.

It had been a strange morning, to say the least, of the kind that could happen only in India. I have loved the country ever since I first went there as a student aged 19, and think I would be perfectly happy to live there, though I recognise that what attracts me about it repels others. For me, it is the most profoundly human place on earth, the glory and desolation of human existence being constantly before one there in a way that is matched nowhere else.

One of the strangest things about the episode I have just related, however, was that at no moment did I ever feel frightened, at risk or in danger. I knew perfectly that India sometimes erupted into the most hideous intercommunal violence, but in between times it seemed more peaceable than almost anywhere else on earth, certainly anywhere else on earth with such large conglomerations of people living in proximity to one another. The obvious fact of my (relatively) enormous wealth in the midst of such obvious impoverishment aroused no hostility in the slum-dwellers and no fear in me.

Since then, of course, Bombay has become much richer, more populous and more dangerous. (In my day, India was still in the throes of Nehruvian socialism, no doubt economically disastrous but, dare I say it, preservative of a great deal of the charm of the country.) But the latest outbreak of terrorism eclipses all that has gone before.

It is never long, when a group of terrorists behave in this hideous fashion, before someone finds, if not quite a justification for their actions, at least an explanation that slides or slithers into such a justification. And so it was in the British liberal newspaper that a well-known and distinguished writer on Indian history and affairs, William Dalrymple, wrote an article with a heading (not chosen by him, in all probability), 'India's poor human rights record in [Kashmir] has ignited the wrath of a new brand of terrorist - well-educated and middle-class.'

This heading accurately conveys the message of the article: that well-educated, middle-class terrorists (one of them wearing a Versace tee-shirt) were so angered by the situation in Kashmir, and India's determination to hang on to the territory, as well as elsewhere, that they resorted to mass killing in Bombay. The article says of the terrorists that:

> These were not poor, madrasah-educated Pakistanis from
> the villages, brainwashed by mullahs, but angry and well-ed-
> ucated, middle-class kids furious at the gross injustice they
> perceive being done to Muslims by Israel, the US, the UK
> and in India in Palestine, Iraq, Afghanistan and Kashmir re-
> spectively.

It would take an entire book, perhaps, to disentangle all the assumptions and misconceptions that this passage implies, or on whose connotations it depends for its force.

Let me refer first to the surprise that it should be educated, middle-class young men who perpetrated such acts. The assumption underlying this surprise is that there is some direct connection between poverty and ignorance on the one hand, and extreme political violence or terrorism on the other. Well-to-do people are not driven to the desperation of terrorism. And this view, it seems to me, genuinely implies an almost total absence of knowledge of world history, to say nothing of an inability to make fairly obvious connections.

Although I am not an historian, it has long seemed to me that some acquaintance with the history of Nineteenth Century Russia is absolutely crucial to understanding the modern world, for it was there that the various forms of modern revolutionary terrorism, and politics as the pursuit of an ideological end, first developed. And the first terrorists were certainly not downtrodden peasants brainwashed by religious or other leaders: they were either aristocrats suffering angst at their own privilege in the midst of poverty, or members of the newly-emerged middle classes, angry that their education had not resulted in the influence in society to which they thought themselves entitled by virtue of their intelligence, idealism and knowledge.

This pattern has been repeated over and over again. Latin America is a very good example. Castro was the spoilt son of a self-made millionaire who had a personal grudge against society because he was illegitimate and sometimes humiliated for it; in other words, he was both highly privileged, with a sense of entitlement, and deeply resentful, always a dreadful combination. Ernesto Guevara was of partially aristocratic descent, whose upbringing was that of a bohemian bourgeois, who was too egotistical and lacking in compassion for individual human beings to accept the humdrum discipline of medical practice.

The leaders of the guerrilla movement in Guatemala (a country, oddly, with many parallels to Nineteenth Century Russia) were of bourgeois and educated origin; one of them was the son of a Nobel-prize winner, not exactly a true social representative of the population. The leader and founder of Sendero Luminoso of Peru, a movement of the Pol Pot tendency (and Pol Pot himself, of course, studied in Paris), was a professor of philosophy, and his followers were the first educated generation of the peasantry, not the peasants themselves. Peasants are capable of uprisings, no doubt, even very bloody ones, but they do not elaborate ideologies or undergo training for attacks on distant targets.

Let us now take the supposed anger at the injustices or human rights abuses committed against Muslims worldwide by various countries, including India.

I do not want to imply that no one is capable of being moved to anger by injustices committed against others, or that it is impossible to care deeply about the fate of some section of humanity remote from oneself. It would be cynical, unjust and simply unhistorical to deny that, for example, William Wilberforce was not genuinely moved by humanitarian motives to bring about the abolition of the slave trade.

But Wilberforce did not demonstrate the depth of his feeling or resolve by killing anyone, nor is it possible to imagine him having done so.

How can outrage at the supposed lack of humanity of others, at their violation of human rights, lead to killing people at random?

There are, I suppose, two possible answers to this: that the people killed were not selected at random, and (or) that no other resort was left to the angry young men. But neither of these defences can possibly work, or extenuate what they did by so much as a jot.

They could not argue that in attacking the Taj Hotel and other such targets they were striking at those responsible for the policies and actions that supposedly infuriate them, simply because they were the resort sometimes of rich citizens of the countries that they hated. To argue like this would be to make every Moslem corner shop owner in the north of England guilty of Osama bin Laden's acts, which is both absurd and morally repugnant.

Moreover, you don't have to know much about how grand hotels work to know that if you start spraying machine gun fire in them you are going to kill a lot of poor people as well as some rich ones (somehow, it is always the poor who get killed first).

Therefore, making the justified assumption that the terrorists were not actually deficient in raw intelligence, it was not the target that was important to the young killers, it was the act of killing itself. And their manipulators probably knew that there are always fools enough, at least among intellectuals in the west, to assume that if you go to extreme lengths, you must have some 'cause' - which is to say some good cause - that impels you to go to them.

The second possible justification, that no other resort was left to them, is likewise absurd. The article in *The Observer* that I have cited claims that neither the Indian state nor the Indian press has investigated or publicised human rights abuses in Kashmir, but this is simply not so.

It is not difficult to find articles about such abuses in the Indian press; but even if this were not the case, there is no evidence that the terrorists, who were quite obviously willing to die, had tried anything else before they tried random killing. In their case, they killed not as a last but as a first resort. It was the answer to their need for significance.

It is highly likely, of course, that the young men's immature or adolescent angst was manipulated by older men with a clear and strong, if intellectually nugatory, ideological outlook. That outlook has absolutely nothing to do with the good of humanity, any more than did Lenin's. Indeed, the article in *The Observer* quotes the leader of the Lashkar-e-Taiba movement in Pakistan as saying that, Christians, Jews and Hindus being the enemies of Islam, the aim of the organisation is to 'unfurl the green flag of Islam in Washington, Tel Aviv and New Delhi.'

The significance of this passed him by, because we also learn from that article that 'there is unlikely to be peace in South Asia until the demands of the Kashmiris are in some measure addressed and the swamp of grievance in Srinagar somehow drained.' As the Duke of Wellington replied when a stranger said to him, 'Mr. Jones, I believe': if you can believe that, you can believe anything.

4
The Rules of Perspective

If I were asked, without time to give the question much thought, to name the greatest political virtues, I should reply, 'Prudence and a sense of proportion.' A New Jerusalem could not be built of these virtues, of course, but neither could a Hell on Earth; and since Hells on Earth are two a penny in human history, but New Jerusalems are infrequent, to say the least, there is much to be said in presumption of those admittedly tepid and unexciting qualities.

No doubt prudence is in part a genetic gift, for people vary by heredity in their temperament. Some there are who are born excitable, and some who are born phlegmatic, though experience will generally teach even the most irascible of people to control themselves when it is in their interests to do so. No one is so adventurous that he puts his hand into the same fire twice – as Heraclitus never said.

But a sense of proportion is the more difficult and elusive quality. How is it learned or achieved? Does experience teach it, or reflection upon experience?

Tell me where is fancy bred?
Or in the heart, or in the head?

The same question could be asked of a sense of proportion.

I was impelled to think about this question by a recent book review in the British liberal newspaper, *The Observer*. The book under review was *Hitler's Private Library: The Books that Shaped His Life*, by Timothy W. Ryback. No sooner had one read the title than one wanted to read the book. Hitler is one of the very few figures in history of whom, somehow, one can never have enough biographical information, as if, were we to pluck out the heart of his mystery, we would simultaneously have plucked out the heart of the mystery of human evil. And the books that he read, and the annotations he made, hold out the promise, or at least the hope, of some new insight into his character. No doubt this is illusory (for very obvious reasons), but I shall still buy the book. Indeed, I await its arrival with excitement.

The review begins in unexceptional fashion:

Dictators tend to be night workers, immune to the exhaustion that topples the rest of us. Napoleon, in an official portrait by David, posed in a study with closed curtains and a clock marking 4 am; as if on sentry duty, the vigilant emperor oversees his dormant, submissive realm.

Two lines further down, we read:

The sergeant who issued orders to Corporal Hitler in the trenches in 1915 was impressed by his insomniac underling, who even then seemed – at least in the officer's obsequious recollection – to be destined for greatness.

But what of the two lines squeezed between Napoleon's and Hitler's insomnia (or, really, in the case of Hitler, his sleep reversal)? What third example – for examples must always come in threes, like serious accidents such as plane crashes, for rhetorical effect – are we offered?

If you are now trying to think of sleepless dictators who came between Napoleon and Hitler, give up now. The reviewer is nothing if not anachronistic:

Margaret Thatcher boasted of making do with an hour or two of sleep.

This is no mere slip of the pen; for, having described how Hitler contemptuously claimed to the abovementioned sergeant that sleep was for others, and did not matter to him, the author of the review continues:

> But such hyperactive despots have a problem. What can you do during those white nights, with the rest of your government peacefully snoring?

So Mrs Thatcher is a despot to be uttered in the same breath as Napoleon and Hitler. (Actually, I don't think Napoleon and Hitler are to be mentioned in the same breath, psychopathically indifferent as Napoleon might have been to the deaths of hundreds of thousands of soldiers.) There could not be a clearer instance of a lack of that sense of proportion that I have extolled as among the greatest of political virtues.

Here, let it be clear, I do not speak as an apologist for or uncritical admirer of Mrs Thatcher. Her achievements are likely to look more considerable to foreigners than they are to native Britons. Her effect on her own society was equivocal, to say the least: some things she did were good, but some dreadful, in their long-term effects. Perhaps this is hardly surprising. It is not in the nature of practical politics to bring about unequivocal benefits, and politicians are men (and women), not gods.

Nevertheless, I do not see how any reasonably sensible person, with pretensions to intellectual balance such as a book reviewer is supposed to have, could put Mrs Thatcher in the Napoleon class of despot, let alone the Hitler class. No doubt she was brusque with her own colleagues sometimes, but this is not the same as leading an army of 600,000 men into Russia and returning with at best a few thousand of them, apparently personally quite unchastened by the whole experience.

How does such a grotesquely unjust, absurd and morally frivolous comparison come to be made, by an author who is in all probability proud of having made it?

There were undoubtedly many thousands, indeed millions, of people in Britain who were harmed, at least in the short-term, by the so-called Thatcher revolution. Mrs Thatcher was determined that never again would a British government be held to ransom by the coal miners' union; and so she set about destroying the latter by means of closing down the coal-mining industry altogether. Environmentalists, I suppose, can only applaud this decision.

There were many towns in Britain that were wholly dependent economically upon coal mining; and since they had been dependent upon it for upwards of a century, a strong, and in many respects admirable, sub-culture had grown up in those towns. This sub-culture that went along with the hard, dangerous and unhealthy life of coal-mining was destroyed utterly by Mrs Thatcher's reforms, and many of the towns in question became sinks of despair, in which the only alternative to unemployment seemed to be sickness. In the town of Merthyr Tydfil, for example, in South Wales, twenty-five per cent of the adult population claims to be too ill to work rather than merely unemployed.

Now if some person from a town like Merthyr Tydfil had placed Mrs Thatcher in the same category as Napoleon or Hitler, one could understand it, and perhaps even sympathise with it a little as an expression of understandable despair. (Few people, mind you, can work up the same rage at the leaders of the Miner's Union, that provoked a suicidal showdown with Mrs Thatcher, and were at least as responsible as she for the closing of the mines.)

But however much one sympathised with the person making the comparison, it would still be wrong, if it were meant seriously, if it were meant to imply that Mrs Thatcher was as morally monstrous as Hitler. The man making the comparison is unable to see that his own grievance, even if justified, and very real as it is to him, cannot be sufficient grounds for such a comparison.

It is a fair bet, of course, that the author of the review that included Mrs Thatcher in the class of maniacal despots did not suffer at all during her period of office, unless hatred of her be regarded as a form of suffering. Indeed, it is far more likely (though I do not know this for certain) that the author of the review was one of the class of beneficiaries of her reforms.

His lack of a sense of perspective about her therefore does not even have the experience of personal suffering caused by her to extenuate it. Indeed, it seems to me more likely that the lack of a sense of perspective exhibited in the review derives precisely from the lack of suffering ever experienced by the author, at least as the result of living under a despotism, as well as a failure of the imagination. A man who could seriously compare Mrs Thatcher with Hitler has clearly made no serious effort to enter imaginatively into life under the Nazi regime.

This failure is by no means uncommon. Virginia Woolf exhibited it just before the Second World War. She had little excuse, for she had travelled in Nazi Germany with her husband Leonard, himself a Jew. Nevertheless, in her book *Three Guineas*, she systematically failed to notice a difference between the Britain of her day, unsatisfactory in many ways as it no doubt was, and Nazi Germany. Indeed, she saw no difference, or none worth remarking upon, and in her book she sided decisively with book-burners (though she wanted to burn different books from those the Nazis burnt). She was misled into such foolishness, I think, by her inability to distinguish between the occasion of her own dissatisfactions on the one hand, and manifest evil on the other.

Another prominent author who made a not dissimilar mistake was Salman Rushdie who, shortly before being condemned to death by the late Ayatollah, also compared Mrs Thatcher's Britain with Nazi Germany, and therefore Mrs Thatcher implicitly with Hitler. No doubt he has somewhat moderated his position on this matter; but the interesting question is why it should take a death threat for a man who, after all, had a degree in history from one of the greatest universities in the world, to betake some thought? Actually, only a fraction of a second's thought is necessary: I don't have a degree in history, but I do not think I should experience much difficulty in pointing out significant differences between Britain under Mrs Thatcher and Germany under Adolf Hitler.

Not only do many intellectuals appear to have difficulty in distinguishing between their own little psychodramas and great world events, so that leaders for whom (sometimes justifiably, sometimes for petty personal reasons) they conceive a dislike become in their minds equal to the greatest monsters in history, but they are inclined to mistake the vehemence of their denunciations for good and sufficient arguments.

The author of the review disliked Mrs Thatcher intensely; he therefore compared her with despots, indeed with one of the worst despots who ever lived; and this, in turn, constituted for him evidence of just how bad she was. Otherwise he wouldn't have compared her with despots in the first place.

Of course, it is possible that there was a simpler error of logic at play:

All despots sleep short hours.
Mrs Thatcher slept short hours.
Therefore, Mrs Thatcher was a despot.

But if so, the error was probably the consequence of the strength of the desire to reach a fore-ordained conclusion.

I do not exclude myself from the temptation to lose sight of perspective. When I work myself up into a pleasant lather of indignation about something in the world (pleasant, that is, for me), which happens not infrequently, I cannot claim always to place it judiciously on a scale of human misfortune.

To complicate matters further, it is sometimes – no, often - to the benefit of civilised life that people get things out of proportion. If no one overestimated the moral importance in our society of cruelty to dogs, for example, there would be no people to rescue dogs from such cruelty, and society would be a little the worse for it.

5

Beauty and the Beast

I t is more difficult to write interestingly of good people than of bad; villains are generally more memorable than heroes. A newspaper that reported only acts of kindness and generosity would be insufferably boring and would go bankrupt even faster than those that relay only disaster caused by defalcation. To adapt very slightly Tolstoy's famous aphorism, good people are all good in the same way, but bad people are all bad in their own way.

To write of good people is often to sound either naïve or priggish; whereas to write of the bad is to appear worldly and sophisticated. One of the reasons, of course, for the difficulty of writing interestingly of the good is that there seems so much less to say of them than of the bad. The good act according to principle, and are therefore lamentably (from the literary point of view) predictable. Once you know how they behave in one situation, you know how they will behave in others. The bad, by contrast, have no principles beyond the pursuit of short-term self-interest, and sometimes not even that. They are therefore unpredictable and their conduct is infinitely various. As I discovered in my medical work, the variety of human self-destruction is, like the making of books, without end; and even the least imaginative and inventive may discover new ways of exercising malignity. Since variety is the spice of prose, the bad are lingered upon with affection by most, if not by all, writers.

At the same time, of course, there is the problem of evil: how it arises, and how it triumphs. No one troubles himself to anything like the same extent over the problem of good: how it arises, or how it triumphs. Perhaps this is testimony to the victory of Rousseau's idea that we are fundamentally good by nature, though deformed by society, over that of Original Sin, which proposes that we are all sinful from birth. Suffice it to say that no one would nowadays subscribe to the idea of one of the Wesleys (I forget which) with regard to the beating of children, that it is never too soon to begin God's glorious work.

It occurred to me in view of the problem of good – I mean the literary problem, not the metaphysical one – to try to write interestingly of some of the very good people whom it has been my fortune to encounter in my passage through this vale of tears we call the world. Somerset Maugham once tried the experiment, but I am not sure that he succeeded. It is difficult for wasps suddenly to turn cuddly.

The first person I want to describe (and to whom I want to pay tribute) is the physician for whom I worked immediately after qualification as a doctor. In those days it was not as common as it is now for women to have reached the peak of their profession, and this woman had done so by the sacrifice, how willing or with what inner conflict I cannot say, of everything else in her life. She was what used to be known as 'an old maid,' and though well into her fifties, which seemed to me then a great age, still lived with her mother. Her background was obviously that of the upper middle class; I should guess her family had been in what were called 'easy circumstances,' that is to say in unostentatious but secure financial comfort going back generations.

She was an extremely competent and knowledgeable doctor, though very modest, and in her younger days had done distinguished research. But her most distinguishing feature was her unstinting devotion to the welfare of her patients, whom she always treated with the greatest kindness that not even the roughest of them could fail to notice. I considered them fortunate to have her as their physician, for even when she could do nothing for them her evident love for them reduced their suffering greatly, and implicitly recalled them to their duty to think of others even in extremis, which likewise reduced their suffering. Her influence on others was therefore great, but indirect. If she had a fault, it was to go on wrestling with death too long on her patients' behalf; but when I think of the reality of the possibility of compassion, against those cynics who say that it is really something else, a disguised form of self-interest for example, it is of her that I think, between thirty and forty years ago.

Alas, not very long after I left the hospital, she was diagnosed with a rapidly-progressive form of breast cancer and died, survived by her aged mother. I still cannot think of this without being seized by sorrow. If her life was in any way unfulfilled, as it may well have been, she was too lacking in egotism to obtrude her troubles on others; I imagine a tragic resignation on her part, but perhaps I am projecting myself unjustifiably into her situation. At any rate, she sought neither fame nor fortune, but to do good in a quiet, unobtrusive way, and died genuinely lamented by all who knew her.

At the hospital also was a young Indian doctor who, being a few years older than I at the time, appeared to me to be middle aged. He was a chain-smoker, which in those days did not seem to be such a strange thing to be; no doubt it was a cause of his very early death from a heart attack.

His background was such that he should have been a spoilt brat rather than a man who inspired instant affection, again even from the roughest patients who might in other circumstances have been expected to be derogatory on racial grounds about Indians in general. His personal vulnerability – which had nothing to do with his knowledge or competence, which in fact were considerable – inspired a protectiveness in others, who therefore wished him only well. I am sure that his patients wanted to get better for his sake as much as for their own; they did not want to disappoint him.

He came from a rich family: so rich that when he, a cricket fanatic, wanted desperately at the age of ten to watch a match a thousand miles away, his father would organise a special train and pack him off in the care of assorted servants to watch it. Until he arrived in Britain, like many Indians of the upper or middle classes, he had never so much as carried a case for himself; contrary to what many suppose, such Indians experience a precipitous decline in their standard of living when they reach Europe or America.

I see him now, short and slightly overweight, a man used to the utmost luxury, trudging round the dreary hospital in his white coat at two and three in the morning, treating all manner of emergencies calmly and with immense good humour. His gravelly laughter rings still in my ears; I hear his accent also, his replacement of v by w, and vice versa.

Whenever I have reason to think of the incalculability and essential mystery of mankind, I think of him; for there was everything in his background to make him an egoist, a man who considered his own whim law, yet he turned out to be a selfless, modest, amusing (because

36

always amused) man.

In this connection, I think also of the adolescent son of a female alcoholic patient, nasty and violent in drink, whom I expected to be adversely affected by growing up in an atmosphere of every conceivable kind of squalor, physical, emotional and moral. If he had been truculent and aggressive I should have understood it; if he had thought he was hard done by, I could hardly have disagreed with him. But instead of being such a young man, he was extremely well-mannered and attentive to his own education, not resentful in the least; moreover, he looked after his disagreeable mother with a tenderness that was amazing to behold and (frankly) impossible to understand, considering the life she had led him. Where did such goodness come from? It was at least as difficult a problem as that of evil.

Having not long qualified, I went to work in a large hospital in Africa – Rhodesia to be exact. The hospital was a very good one, and the nurses of one of the wards were supervised by a senior nurse who was fat, jolly, competent and of surpassing kindness. She, too, lived with her aged mother, in what used to be known to the whites as the African township. This consisted of thousands of identical small concrete houses with tin roofs.

The township was dangerous: full of *shebeens*, where vast quantities of maize beer were drunk by young men who became quarrelsome and violent, inflamed by every kind of dissatisfaction and frustration. But the nurse's home, to which I was invited several times, was a haven of tranquillity, spotlessly clean and with a tiny but immaculately kept garden.

I suppose an aesthetic snob (such as I was at the time) would have said that the interior, with its plush chairs with frilly antimacassars, was a petit bourgeois paradise; but in the context of that time and place it was a triumph of the human spirit over great adversity, a considerable achievement. I learnt (I hope) not to despise the decent ambitions of the humble, but rather to see in them the work of civilisation: something that intellectuals are very much inclined not to do, but rather to indulge in demeaning jokes about them, *de haut en bas*.

Quite a number of years later, I met in a remote part of Nigeria an aged Irish nun, well into her seventies, living in an isolated convent with other nuns, who made it her work to bring food to the prisoners in the local prison. I have very little doubt that they would have been severely malnourished or even starved to death without her arduous attentions; she made sure that each of the prisoners, some of whose sentences had expired but who had not the requisite money to bribe the gaolers to release them, and others of whom had been on remand for ten years, was fed. For it was a matter of fact, accepted as a law of nature, that officialdom would steal whatever there was to be stolen.

The nun had nothing but her moral authority to effect her work, and she had no reward but the gratitude of the prisoners and the compliance of the guards. It was clear that they all now had both a respect and an affection for her; she carried around with her an aura of invulnerability to the world's evil. But none of this had gone to her head, on the contrary; her humility was genuine and unselfconscious, and I suppose if asked she would have denied any special merit in her conduct. The reproach to one's own comparative lack of humanity was implicit rather than explicit. The power of example is that it is exemplary, not declarative, much less declamatory.

It is not of course for me to say whether I have been able to interest the reader in some of the remarkably good people whom I have met, or whether they would really rather have heard about the baby-sitter whom I met who killed the three infants in his charge because he didn't like the noise they were making that interfered with his concentration on television. It might be said that, having described the goodness of these five people, I would have nothing more of interest to say about them; whereas, had I chosen the four or five greatest moral monsters whom I had encountered, I would have much more to say.

But this is not quite right; the fact is that we are much more interested in the life histories of the moral monsters than in those of people like the five exemplars whom I have described. Their lives were neither uniform nor without interest, but I did not enquire into them with the same curiosity that I have employed in the cases of the moral monsters.

In summary, it may be said that evil attracts and engrosses us in a way that good rarely does.

6
Attitude or Gratitude?

A recent Dutch visitor to my house in France was observant enough to notice that I disliked wasting food. He told me that he was very much of the same mind.

It occurred to me then to try to find the cause and justification of our dislike of such waste. Where did it, this dislike, come from? What reason could we give for it? (These are not the same questions, of course.)

The Dutch are famously parsimonious, but parsimony is neither one of my vices nor one of my virtues – and I leave it to others to decide which of them it would be if I had it. And I also knew from experience that my visitor did not partake of this national characteristic, if it is one, so that we might safely leave Calvinism to one side. As it happens, we were both children of the post-war era, when material life in Europe was much less abundant than it is now. I remember the days when butter was treated as a luxury rather than as an item produced, thanks to a combination of subsidy and technical advance, in such mountainous quantities that it could, if melted, replace the seas.

Chicken was still a luxury food during my early childhood, even in middle-class households, and our parents referred constantly to the real shortages of the war years. They knew (or so they claimed) how to make a vast omelette from a single egg, suitably expanded by various contrivances. We were led to suppose, therefore, that we were fortunate indeed to have been born in an age without privations.

We were not aware of any actual shortage because we had no standard of comparison, and we always had more than enough to eat. The abundance – one might even call it the superabundance - that was to arrive in the near future was necessarily unknown to us and therefore could not serve as a comparator. If being of the immediate post-war generation had any effect upon us at all, it was to give us a faint and subliminal awareness that it was possible, in certain unlikely circumstances, for material goods to become less rather than more abundant, and that material scarcity was a possibility, if only a remote one.

Still, I should like to think that our aversion to waste, particularly of food, goes a little deeper. It is not only that we disliked wasting food ourselves; we did not like to see others waste it either. It was not therefore a matter of mere personal economy, or fear for the future. I read somewhere that we – that is to say, members of western societies - throw away a third of the food that we buy, and this appalled me.

My dislike of waste does not arise from any appreciation of the ecological need to preserve, heal or (worse still) save the planet. I wish the planet, as I wish humanity, no harm, but find it too large and nebulous an entity to have any genuine feelings towards it: Gaia means nothing to me. And even if it could be proved that wastage was exceedingly good for the planet, I still should not like it.

Nor does my dislike arise from the fact that there are still people in the world with not enough to eat. Occasionally in my childhood, an adult would tell me to eat up (when I was already full) because there were children in Africa who were hungry. This, in those days, did not seem to me logical, since I could not see how eating more than I wanted was any more likely to help the hungry African children than throwing food away. As every child knows, mere logic does not persuade adults: the time I tried the argument out on my mother, she only told me to do as I was told.

After a little reflection, I came to the conclusion that my dislike of waste arises from a whole approach to life that seems to me crude and wretched. For unthinking waste – and waste on our scale must be unthinking – implies a taking-for-granted, a failure to appreciate: not so much a disenchantment with the world as a failure to be enchanted by it in the first place. To consume without appreciation (which is what waste means) is analogous to the fault of which Sherlock Holmes accused Doctor Watson in *A Scandal in Bohemia*: You see, but you do not observe.

It is strange that I had not really thought about why I disliked waste, and of its destructive effect on the human personality, before my Dutch visitor made his observation. But then I can hardly claim to have been an exemplar throughout my life in the matter of waste. I most certainly cannot claim never to have wasted anything, quite the contrary. Paradoxically, perhaps, it is only as I came to be in relatively easy circumstances that waste came to seem so objectionable to me.

This happened at roughly the same time as still life came to be among my favourite artistic genres. Early in my life, I could hardly see the point of them; only comparatively late in my life did I see them as a call to contemplate what exists without taking it for granted, thereby increasing one's appreciation and enjoyment of the world.

Once you become aware of a phenomenon such as waste that you overlooked or considered unimportant, you begin to see – or rather, observe - it everywhere. For example, yesterday I was walking in a street in England and I saw a box of cakes thrown on the ground. One had been half-eaten, but the rest were strewn around, so it was not merely that the box had been dropped by accident. The person who had dropped it had eaten a little and decided that the rest was surplus to his requirements.

I pass over – but not because I haven't noticed it – the unsocial and egotistical way in which the person disposed of what he no longer wanted. Rather, I refer to the fact that whoever disposed of the cakes in that way took them entirely for granted, gave no thought to the effort or ingenuity required to produce them, assumed that there would always be more when and where he wanted them, and in general evinced no respect for anything except his whim of the moment.

I do not have a consistent philosophy of waste and its avoidance to offer, however. I have no desire to be a desert anchorite, living on as little as possible. I like, though I do not crave, luxury. I do not eat only to keep body and soul together, and like expensive foods, though also delight to eat well for little money. I buy books that I could just as well take from the library, and the catalogues of antiquarian booksellers delight me.

I am aware that our whole economic system depends to a large extent upon us consuming vastly beyond our needs, biologically considered, and that if we were all as parsimonious as possible and never threw anything away that was remotely usable or re-usable, the wheels of commerce would soon grind to a halt. I am aware that our prosperity and comfort depends upon those wheels continuing to turn, more or less ceaselessly, and without any higher purpose; but I have no vocation for discomfort or poverty, and suspect that concern for the environment, in so far as it really exists, would melt away faster than the glaciers or the polar ice cap at the first prolonged power cut. When one considers how much fuss people are now inclined to make when something in a hotel (for example) does not come up to the standards of comfort they have come to expect, I do not think a mass conversion to asceticism is on the cards.

Like many social phenomena, abundance is both good and bad. When I was a child, my mother used to darn our socks. I still remember the wooden mushroom that she would insinuate into a sock with a hole, the better to expose the latter for her to close up with wool or cotton thread.

This is now as unthinkable a ceremony as touching for the king's evil (scrofula) would be. Now if we have a hole in a sock we throw it away at once; and if we are short of socks, we go and buy ten pairs for what it takes us two minutes to earn.

I have no real vocation for darning socks; I think I have better things to do with my time (though, truth to tell, I am not entirely sure if this applies to everyone). Attention to and gratitude for socks is not a commonly expressed attitude. And yet I cannot help but think that this habit of throwing things away the moment they become defective leads to an unpleasantly disabused attitude to life. Computers, washing machines, televisions, refrigerators, clothes, out they all go the moment they break down or require repair. I know it is a tribute to our immense productivity that it is far cheaper to obtain a new machine than to repair the old, but in a world where everything is so instantly replaceable, what affection or gratitude develops for anything? What do we notice and appreciate if everything is instantly replaceable?

Not long ago, my wife had a slight car accident - entirely her fault, but I did not reproach her, as secretly I wanted to, since my own driving is not always completely above reproach. The car was eminently reparable, and indeed still functioned perfectly well. But the insurance company insisted on scrapping it, because the repairs would cost more than the car was worth. (I should perhaps mention that I was not entirely convinced of the honesty either of the car repairer's estimate of costs to repair the car or of the insurance assessment of the car's value; I suspected, but could not prove, collusion and skulduggery. Here, if anywhere, a man is completely the victim of forces beyond his control.)

If the car really was scrapped, it offends my sense of waste, though not of economic rationality. The high cost of labour to repair it means, presumably, that the people who repair cars can live at a decent standard of living. Countries in which old cars never die tend to be poor ones.

I suppose that what I would like is an abundance that everyone appreciated and did not take for granted. This would require that everyone was aware that things could be different from how they actually are, an awareness that it is increasingly difficult to achieve. I myself can hardly remember what it was like to live without personal computers and the internet, though I have lived the majority of my life without them. I now take them sufficiently for granted that if, for any reason, I am out of range of the internet, I regard this as something of an outrage.

I still have vestiges of the requisite awareness, however. In my long distant childhood, I had an uncle who was a prisoner-of-war of the Japanese, and I remember how impressed I was when I was told (*sotto voce*) that he still woke up in the night with nightmares of his captivity. He had gone without food, of course, and suffered beri-beri; and to this day I cannot look at rice on my plate without thinking of him. It helps me to look on each single grain as something not to be despised.

In general, a life of assumed abundance is one of ingratitude; one is not grateful for anything that could be no different from how it is. So perhaps when my mother told me that I should think of the children in Africa who did not have enough to eat, and eat up what was on my plate, she was not so much trying to benefit the children in Africa, as to benefit me: to make me grateful, and not to take for granted what, in fact, would almost certainly always be there, namely an ample sufficiency. Without gratitude, there is no happiness.

7
Poisoned by Celebrity

C an someone who appears to have been born without a moral
   sense, or never to have developed one, properly be called 'evil,'
or even 'bad?'

For when we evaluate people and classify them morally, as inevitably we do, whatever those who claim to hate the sin but love the sinner may say, we assume that they are moral agents, able to behave differently from how they actually do behave.

In my work as a doctor in a prison, I came across a number of poisoners, who tended to be more interesting as a group than other murderers. Most murders either take place impulsively, in the midst of a sordid and often drunken quarrel, or in the course of a robbery, or are the settling of scores between criminals, when the victim is scarcely better as a person than the perpetrator, who merely got in first.

But poisoners are different; they kill with more deliberation than others, with more malice aforethought. Not all poisoners have the same motives, of course, though desire for money has played a large part in most of the cases that I have known. Those who did it for life insurance were seldom able to wait more than a couple of weeks after increasing the sum assured by ten times before they struck. A peculiarly dangerous period for old men is shortly after they have changed their will in favour of a favourite nurse. It does not seem to occur to these poisoners that the new financial arrangements will give a rather easy clue as to motive to even the least deductive police investigators.

A man called Graham Young poisoned several people, some to death and others only to near-death, in the 1960s and 70s in England without any pecuniary motive, indeed without any obvious motive at all, starting when he was thirteen or fourteen years of age. Among his victims (who did not die) were his father and his sister. It is probable that he poisoned his step-mother (who was devoted to him) to death.

His sister, Winifred Young, who was eight years older than he, and whom he had tried to poison, wrote a book about him, published in 1973 entitled *Obsessive Poisoner*. It is a remarkable book in several ways, and not only because there can have been few memoirs by people who survived attempted poisonings. It is a valuable document of social history, for it implicitly records a period of great cultural change, not only in Britain but I suspect throughout the world.

Graham Young was born in 1947, to parents of the aspiring lower middle class. His mother died of tuberculosis when he was very young, and his father re-married in 1950, to the woman whom he was almost certainly to poison to death nearly twelve years later.

From an early age, Graham Young showed marked peculiarities. He did not make friends easily, or at all, and in so far as he sought out the company of other human beings it was of people considerably older than himself. He was almost emotionless, apart from a love of dogs. He was highly intelligent and looked down on people who were less intelligent than he, which was most people, but, while good at his schoolwork, was not perseverent in subjects that did not interest him. From about the age of eleven he displayed an obsessive interest in two subjects: the Nazis and poisons. He talked about them incessantly. One of his great heroes was Dr William Palmer, who was known as the Prince of Poisoners, who is suspected of having poisoned a great many of his close relations and friends in the 1850s for financial reasons.

He put belladonna in his sister's tea in 1961, which she was prepared, out of an inability to imagine evil of her brother, to believe was an accident.

Their stepmother died in 1962 of symptoms that, in retrospect, were compatible with thallium poisoning. He put thallium in a sandwich that her brother-in-law ate at the post-funeral collation.

At the same time, Graham Young administered antimony to his father and to a friend of his, so-called, at school, as well as his 'favourite' aunt. With horrible cunning, he sought to console them as they suffered from what he had given them, as he had consoled his dying stepmother.

The penny finally dropped, he was arrested, tried aged 14 for attempted murder (unfortunately, his stepmother had been cremated and her ashes were not available for forensic examination, so he could not be charged with murder itself), and sent – as a psychopath – to Broadmoor, an institution for the criminally insane.

What shines through his sister's narrative is the complete absence of motive for his crimes, and indeed the ordinary and even banal goodness of everyone by whom he was surrounded. Whatever else might be said, it could certainly not be said that his background had anything to do with what he did. His father was a steady, hardworking man, without obvious character defects, not very interesting or exciting perhaps, who for years did overtime in order to pay off the mortgage on his house - which he did, in fact, in 12 years - to secure the future of his children. He was the very archetype of the reliable, modest, industrious, law-abiding citizen upon whom the maintenance of civilisation partly, but importantly, depends.

Furthermore, the goodness of the author herself is obvious, precisely because she is herself so unaware of it. Not only was she reluctant to believe evil of her brother, but even when that evil became manifest to her she did not cast him into outer darkness. Her love, the ordinary love of a sister for a brother, exceeded her condemnation of him: which did not mean, however, that she sought any legal exculpation for him, or made any excuses for him. She loved him as a brother, but as a citizen she knew that he had to be punished and the public protected from him.

It is worth pointing out here that this morally sophisticated attitude was not that of an exceptionally-educated person: she was a secretary, without tertiary education. In other words, her moral sophistication was absorbed from the general culture, not from explicit teaching.

(Incidentally, but not coincidentally, her book was extremely well-written, far, far better written than many people with postgraduate degrees could write such a book today).

Graham Young spent ten years in Broadmoor, before being released in 1972. Between the time of his arrival and his departure the whole ethos of society had changed. He arrived shortly after a man died there in his eighties, having been sent to the institution (for a crime which could hardly have been more serious than Young's) more than seventy years before. But, despite the fact that two psychiatrists at the time of his trial had asserted that it was unlikely that his perverse interest in poisons would ever decline, he persuaded his psychiatrist at Broadmoor, by then probably seeing himself in the role of Graham's St

George against the dragon of society, that he was 'cured.' And this, despite the fact that a patient at the institution in the meantime had almost certainly been poisoned to death with cyanide distilled from laurel leaves in the hospital grounds (though admittedly, this had not been proved to be Young's handiwork) and that – beyond doubt – he had attempted to poison the tea of nearly a hundred fellow-inmates. By now, it seems, the need to think well of humanity in general trumped altogether the disinterested and objective examination of particular instances of it.

So Graham Young was released. He was sent to a government rehabilitation centre. Within two months, he was buying dangerous poisons again. But no one who dealt with him was informed of his background or his previous history, not even the probation officers whom he was told to visit every two weeks. Was he not cured? Did not the director of Broadmoor say that, if they thought there was any risk at all, they would not have released him in the first place? It would therefore be unfair to him, unduly prejudicial, to let people – anyone – know what he had been up to all those years ago.

A job was found for Graham Young in a photographic factory. His employers knew that he had had 'mental problems' that accounted for his lack of an employment history, but they took the laudably unprejudiced view that everyone deserved a chance, and that no one's past should be held against him. The employers were not told that he had been a poisoner or an inmate of an institute for the criminally insane for ten years, and so when members of their staff began to suffer mysterious symptoms shortly after his arrival, they did not connect him with them.

So many of the staff, in fact, began to suffer from such symptoms as nausea and polyneuritis that the public health authorities were called in. The most likely explanation seemed to be a virus, especially as the factory was searched high and low for heavy metals that might equally have caused the symptoms, and none was found. Two of the staff died, and it was largely because Graham Young himself asked the local doctors at a meeting that they convened in the factory whether they did not think that the illness from which the deceased had died might not be thallium poisoning that he was first suspected and then arrested.

Stores clerks are not normally expected to be knowledgeable about toxicology, but so eager was he to prove his superiority over the doctors in public that he over-reached himself. Phials of thallium were found among his possessions in his lodgings.

Two things struck me about the narrative, apart from the literacy and goodness of his sister. The first was the deeply old-fashioned stoicism and devotion to duty of the staff of the company for which he worked.

Many of them were made desperately ill by his addition of poison to their tea (he again used two poisons, as he had when he was fourteen years old, antimony and thallium which, because they caused rather different symptomatology, confused the public health doctors), but despite being hardly able to walk or to hold anything down, they insisted that they would soon be all right, and continued to try to work. Above all, they did not want to make a fuss, until some of them were admitted as emergencies to hospital.

The other thing that struck me was the obvious and sometimes openly expressed desire of Graham Young to achieve celebrity by his poisonings. He wanted to be known and remembered as the greatest poisoner in history; he took pleasure in the publicity that he received, and he was more concerned with the newspaper coverage of his first trial than with the medical condition of his blameless father whom he had poisoned.

He lived at a time of a fundamental shift in our culture. On the one hand he was very old-fashioned; he dressed conservatively, always in a shirt and tie, and with a handkerchief in the breast pocket of his suit. In writing to his future employers to accept the job they had offered him, he ended his letter as follows:

> I shall endeavour to justify your faith in me by performing
> my duties in an efficient and competent manner.
>
> Until Monday morning, I am,
>
> Yours faithfully,
>
> Graham Young

This is the language of an era soon to be as bygone as that, say, of the Etruscans.

On the other hand, he matured at the time when the cult of celebrity, for celebrity's sake, was fast gaining ground. It was a new form of celebrity, disconnected from any solid form of achievement, of which an ability to attract publicity became the *sine qua non*. Graham Young was highly intelligent, without the character to stick at anything to achieve something solid, but with a burning desire to be acknowledged as superior, important and outstanding.

When trying to explain why he could not get close to people, he once said to his sister (and she ends her book with these words), 'You see, there's a terrible coldness inside me.' Could a spread of that coldness not help to explain our contemporary preoccupation with celebrity?

8
Steady As She Goes

The relation of language to thought has long been a philosophical puzzle, one to which no universally accepted answer has yet been given. Is language a precondition or determinant of thought, or thought a precondition and determinant of language? For myself, I incline to the latter view, on the no doubt simplistic grounds that, when writing, I often have the following experience.

I know that there is something I want to say, but at first the right words do not come to express it. They are, I realise, only an approximation to my idea; then suddenly, dredged from I know not where (though it feels like somewhere located near the base of my skull), the right words arrive and I know at once that they are the best possible words in my possession for what I want to say.

I suppose it might be argued that somewhere in my preconscious there is a linguistic representation of what I am at first unable to verbalise, and that my little eureka experience (so delightful that it makes the struggle seem worthwhile) is only a recognition that the words in my consciousness now accord perfectly with those in my preconscious. Be that as it may, it seems to me that my experience suggests that conscious thought, at least, can be pre-verbal, even when it is propositional in nature.

Not every one agrees, of course, and in *Nineteen Eighty-Four* Orwell put forward the rather dismal idea that reform of language – that is to say, the imposition of certain locutions and the prohibition of others – can actually mould the content of thought, making some ideas unthinkable and others unchallengeable.

This, of course, is what politically-correct language is all about. It is certainly what its proponents hope.

I was recently the victim of a politically-correct sub-editor of a distinguished medical journal for which I write. I do not claim to have suffered inordinately as a result; at most I experienced a brief spasm of anger, leading to a slightly longer period of irritation. Then I calmed down: 99.99999 per cent of the world's population would never read what I wrote, and of the 0.00001 per cent that did read it, 99.99 per cent would not notice the change.

On the other hand, as Hume said, liberty is seldom lost all at once; usually it is nibbled away, until – to change thinkers to Tocqueville – people become 'a herd of timid and industrious sheep, of which the government is the shepherd.' (It needn't be the government that does all the shepherding, intellectual apparatchiks will do just as well.)

Therefore, at the risk of sounding and even becoming a little paranoid, and of seeing dangers to our freedom lurking everywhere, even in insignificant phenomena, it is necessary sometimes to protest at the most minor acts of arbitrary power.

The paragraph that was altered by the sub-editor of the medical journal went as follows:

> Modesty was once considered a virtue, but nowadays it is clearly an impediment to a successful career. We prefer – or perhaps I should say we demand – boastfulness. To judge by [medical] job applications, the world is now stuffed full of paragons whose moral commitment to the welfare of humankind is equalled only by the brilliance of their contributions to medical science.

Now I should have rejoiced had the sub-editor made two or three alterations to the paragraph – which I wrote very quickly, and did not bother to re-read. I am very far from believing that a sentence is perfect just because it is mine. Had the sub-editor removed the words 'it' and 'clearly' from the first sentence, it would have read better. The sentence would have been more graceful rhythmically and no essential thought would have been lost. And, other things being equal, fewer words are better than more.

Moreover, I don't really like the locution 'stuffed with.' This does not seem to me one that is properly applied to human populations; probably, it is never very elegant. I think 'pullulates with' would have been better.

But these are not the changes the sub-editor made, and on which I would have congratulated him or her had he or she made them. No; the change made by the sub-editor was to substitute 'humankind' for 'mankind.'

My first objection to this is aesthetic. The phrase 'whose moral commitment to the welfare of mankind' is much more elegant than the phrase 'whose moral commitment to the welfare of humankind.' Indeed, the inclusion of an extra syllable ruins the rhythm altogether, and turns it into a horrible mouthful; it sometimes seems to me that sub-editors are chosen exclusively for their cloth-ears and indifference to the beauties of language. Perhaps they are given tests by editors that consist of a choice between two lines of William McGonegall, self-styled Poet and Tragedian of Dundee, and two of, say, Keats:

On one occasion King James the Fifth of Scotland, when alone, in disguise,
Near by the Bridge of Cramond met with a rather disagreeable surprise...

No, no! go not to Lethe, neither twist
Wolf's bane, tight-rooted, for its poisonous wine...

If, by instinct, the applicant for the post of sub-editor chooses the McGonegall, the position is his.

But my other objections to the substitution are more serious, at least if moral considerations are more important than aesthetic ones. Of course, to object to the use of the use of the word 'mankind' because it is sexist is as absurd and literal-minded as to object to the word 'person' because it, too, is sexist: who, after all, is this 'per' whose son has given his name to everyone in the world? Surely, to be absolutely egalitarian between the sexes, it should be 'peroffspring'?

Come to think of it, 'humankind' is also sexist, very nearly as sexist as 'mankind,' for it contains the world 'man.' It should therefore, in all consistency, be changed to 'peroffspringkind.' Moreover, the word 'woman' should likewise be changed to 'woperoffspring.' The possibilities for language reform are almost infinite, at least in English.

Enough of satire – if only because satire these days has an inherent tendency to turn into prophecy. There are enough mad ideas in the world without my adding to them. Let us turn, then, to the meaning of this substitution.

First, it was done without my permission. Now either the sub-editor considered that it was a matter of no importance, in which case there was no point in doing it; or it was a matter of some importance, in which case my permission ought to have been sought. In either case, the change was an exercise in and of raw power, since no writer (at least of my standing) is as powerful as one of the most widely-read medical journals in the world. We writers have two choices: submit to this kind of thing, or shut up.

Was the substitution by itself an example of the descent into Newspeak? It is unlikely that anyone thinks that machismo or misogyny will be much reduced by the universal and compulsory adoption of humankind in place of mankind, let alone that it will actually do so. And yet it is one manifestation of language reform that is intended first to make people afraid to say certain things, then to think them, before reaching the highest stage of thought-control: to make them unthinkable.

Things are worse in this respect, and have gone further, in America than in Britain. Reading American academic books as I quite often do, I have been struck by how common, indeed universal, the use of the impersonal 'she' has become. Occasionally, authors flounder as they try to alternate the impersonal 'she' and 'he' (incidentally, the phrase 'she and he' has now replaced 'he and she,' though to my ear the latter is more euphonious): for quite often when they try it, they do not remember whether their last impersonal pronoun was 'he' or 'she.' Incidentally, I have never heard anyone say 'the hangperson' instead of 'the hangman,' or even 'the taxperson' instead of 'the taxman.'

Now I don't really mind if academic harridans want to insist on using the impersonal 'she;' that is their prerogative and, after all, their words are their own. It is their right to use what words they wish. But I strongly suspect, when I read a book by an elderly and distinguished academic who was brought up, educated and got tenure before the impersonal 'she' was even thought of, and who may very well be the world expert on the subject about which he writes, that its appearance in his books is mandated by the university presses themselves, or by their sub-editors. Here an ideological obsession, cheap and silly as it is, trumps any sense of reverence towards real distinction. This is barbarism.

It is also the worst and most dangerous form of censorship. There are two kinds of censorship, the negative and positive. The negative proscribes, the positive prescribes.

As far as art is concerned, there is a lot, historically, to be said in favour of negative censorship. Most of the greatest art, certainly, has been produced in conditions of such censorship, and – as we have seen in the last thirty or forty years – an absence of even the degree of censorship that can be ascribed to self-restraint has not necessarily resulted in an improvement on the paintings of Piero della Francesca, the plays of Shakespeare or the music of Bach. Of course, there are arguments against negative censorship, but the good of art is certainly not one of them.

The positive kind of censorship is much worse than the negative and, if it goes very far, is almost incompatible with either deep thought or good art. It co-exists with the negative form of censorship, but in addition to making some things unsayable it prescribes what must be said, in the way that any thesis on any subject whatever in the old Soviet Union was obliged to carry quotations from Lenin, showing that Lenin had come to the right conclusions years before. Of course, intelligent people quoted Lenin with satire in their hearts; but forcing men publicly to mouth sentiments as a precondition of furthering their careers is a sovereign way to destroy their probity and induce a state of self-contempt. And men who are contemptuous of themselves are more likely to take to the bottle than to constructive activity.

There is an informal type of positive censorship (at least, I assume it is informal), as well as a formal one. For example, I have noticed of late that when American academics want to illustrate the concept of genius with a list of geniuses, they almost invariably include a sportsman. I doubt that university presses insist upon this as 'house style,' as it were; rather, the academics concerned fear to be accused of elitism by their peers, and elitism carries with it all manner of unpleasant political associations. In a free, or free-ish, society, there is no fear as great as that of losing caste.

The quasi-compulsory inclusion of sportsmen in lists of geniuses is not socially harmless or without effect. To suggest that a basketball player can be compared with Mozart is to put all human activities on the same level; and since some activities come easier and more naturally than others, it has the effect of reducing, indeed making quite pointless, any form of cultural aspiration. If what comes easily is as good as what comes only with deep effort, thought and intelligence, why go to any trouble? There is a Gresham's law of culture: without a scale of values, the bad will always drive out the good.

Thus the intelligent flatter the unintelligent, without, in their hearts, meaning a word of what they say. What seems at first sight progressive is in fact deeply reactionary, in the worst possible sense: it does not permit or encourage social ascent, and changes class societies (such as capitalist democracies have always been) into caste societies.

In the meantime, I look forward (though do not expect) the time when Satan will be referred to routinely in academic books as 'she.' That would be one giant step for humankind.

9
Fujimori

D oes the end justify the means? This question, difficult to answer in the abstract with a categorical negative or affirmative,
occurred to me when I read that Alberto Fujimori, former president of
Peru, had been sentenced to seven and a half years' imprisonment for
corruption, to run concurrently with the twenty-five years he is already
serving for abuse of human rights.

As it happens, I was in Peru just before, during and after the election that first brought Fujimori to power. His opponent was the world-
famous novelist, Mario Vargas Llosa, who I, like many others, assumed
would win. Indeed, I hoped that he would win. He was highly intelligent, extremely eloquent, had a clear idea of what was needed for Peru
to emerge from its current nightmare, and he was standing for election out of patriotism and for the good of his country. He had nothing
to prove, nothing to gain; it is rare indeed to encounter a candidate so
transparently unmotivated by personal goals.

Fujimori won. I hadn't appreciated just how much his obscurity
might help him, so great was the disillusionment in the country with
national figures. Fujimori was a distinguished academic agronomist, but
you could be the most famous agronomist in the world and still live in
the most perfect obscurity. One Peruvian peasant captured the mood
perfectly when asked why he had voted for Fujimori. 'Because I didn't
know anything about him,' he replied. In other words, every man's past

disqualifies him from high public office.

The Peru that Fujimori inherited was in terrible condition. Inflation was so rapid that you couldn't buy anything of any value in the local currency: you had to use dollars. Money-changers, of whom there seemed to be thousands, stood in the streets, waving thick wads of notes at passers-by in exchange for dollars. Once, in Arequipa, my friend and I walked out to visit a convent there. The rate was 90,000 *intis* per dollar (and each *inti* was 1,000,000 old soles) on the way; on the way back, an hour later, it was 110,000 – or, to put it more dramatically, 110,000,000,000 old soles. I suppose that inflation of this kind at least makes you adept at mental arithmetic.

But inflation was, if not the least of Peru's worries, at least not the worst or greatest of Peru's worries. That honour belonged to *Sendero Luminoso* (Shining Path), the Maoist insurgency that at the time controlled quite a lot of the national territory. I was convinced that, if Sendero won, there would be another Cambodia in Peru: a Cambodia on a much larger scale. And it was far from certain at the time that Sendero would not win. Indeed, if I had had to put my money on it winning or losing, I think I would have put it on it winning.

The history of Sendero was instructive, from two points of view. The first is that it destroys the notion that such revolutionary movements are the direct and spontaneous product of the grievances of the poor. The second is that it illustrates the dangerous folly of expanding tertiary education as a means of economic development rather than as a consequence of economic development.

The founder of Sendero was the professor of philosophy at Ayacucho University, Abimael Guzmán, known to his acolytes as Presidente Gonzalo; his ideas, if such they merit being called, being the application of Maoism to Peru, were known collectively as Gonzalo Thought. Although living in clandestinity, he was already the object of a grotesque cult of personality and he wrote and spoke in that terrible *langue de bois* that is not the least of the tortures inflicted on society by communist regimes because it claims a monopoly of public speech and bores into the brain like a loud burrowing insect:

> The ideology of the international proletariat erupted in the crucible of the class struggle, as Marxism, becoming Marxism-Leninism and, subsequently, Marxism-Leninism-Maoism. Thus the all-powerful scientific ideology of the proletariat, all-powerful because it is true, has three stages: 1)

Marxism, 2) Leninism, 3) Maoism; three stages, moments or landmarks of its dialectical process or development; of a single entity that in a hundred and forty years, from the Manifesto, and in the most heroic epoch of the class struggle, in the bloody and fruitful struggles of the two lines within each communist party and in the immense labour of the titans of thought and action that only the proletariat could generate, three inextinguishable luminaries stood out: Marx, Lenin Mao Tse-Tung, who through three leaps have armed us with the invincible Marxism-Leninism-Maoism, today principally Maoism.

Ayacucho University itself had been in abeyance since the seventeenth century; the Peruvian government thought to revive it as a means of developing the economy of the local area, one of the poorest and most backward in the country, and bringing to it a modicum of social progress. What it brought instead was a Peruvian Pol Pot (who had written his thesis on Kant), who was easily able to influence and indoctrinate young men and women who were the first generation ever to receive tertiary education, and who were, in all truth, the scions of an immemorially oppressed people.

The combination of millenarian hopes and age-old resentments is an unfortunate one, to say the least; Gonzalo Thought, so called, gave ideological sanction to brutality, and turned sadistic revenge into the fulfilment of a supposedly scientific destiny. From what I personally saw in Ayacucho on the eve of the election, which had the atmosphere of a city under siege, waiting for the barbarians to arrive and carry out their long-announced massacre, I was convinced that, if Sendero achieved power, millions would be slaughtered.

I also saw, and heard about, actions by the Peruvian army that were less than gentlemanly. People suspected of Senderista sympathies were disappeared (it took the twentieth century to turn the verb 'to disappear' into a transitive one). I saw relatives petitioning the local garrison officer for news of their husbands, sons and brothers whom the army had whisked away and obviously consigned to permanent oblivion. The army did not say please and thank you for what it commandeered; it was more an occupying force than a protector of the people.

Still, it was what stood between Peru and the Apocalypse. But, at the time of Fujimori's election, it looked as if it might collapse.

On my way back to Europe, I happened on the aircraft to sit near a man who turned out to be an investigator for Amnesty International. When I told him about what I had seen the Peruvian Army do, he looked like a man who had just been fed with a tantalisingly delicious dish, or a cat at the cream; it was, it seemed to me, exactly what he wanted to hear. He almost purred. But when I told him what I had seen Sendero do, his expression turned sour; and he looked at me as if I were a credulous bearer of tales about unicorns or sea-monsters. He turned away from me and took no further interest in my conversation. No doubt illogically, I lost a great deal of my respect for Amnesty after that; constituted governments do a lot of evil, but they are not the only ones to do evil. In this case, the government was the lesser evil, and by far.

But it was very weak. The egregious Alan García was the president, and when it came to corruption, and no doubt many other failings, he could have taught Fujimori a thing or two. At the end of his mandate, he sailed away like a Spanish viceroy of old, to enjoy ill-gotten millions.

It was under Fujimori's presidency that Sendero was defeated. The odious and murderous Guzmán was captured, and made to look ridiculous as well as hypocritical. This seemed to me an immense achievement, an uncommon victory over evil.

But, of course, some of the methods used to achieve that victory were not up to the standards of Scandinavian democracy. Years later, after Fujimori had shown an discomforting attachment to power, and the memory of the situation he inherited had faded somewhat, he was charged with having ordered kidnappings and murder, as well as other offences. And indeed, he was guilty of these things.

How does one assess his moral, as against his legal, guilt? Is it permissible to commit a lesser evil to avoid a greater one? I am not a utilitarian, but it seems to me unrealistic to say that we should never depart from the ideal in order to prevent a much greater departure from the ideal; that, like Kant, we should tell a murderer where his victim is simply so that we do not commit the moral fault of telling a lie. On the other hand, the doctrine that the end justifies the means has been responsible for many horrors, large-scale and small.

Let us take the Fujimori case. Our assessment must depend upon things that cannot be known indubitably. For example, can it be known for certain that Sendero might have won the war in Peru? No, it cannot. Such knowledge is radically beyond our powers to attain.

Can it be known for certain that Sendero would have been as bad as I have suggested? The only guide we have is other regimes that have espoused a similar ideology, and they have all committed terrible atrocities, some of the worst in human history. Sendero had so far given every indication of following the pattern; it committed many atrocities even before it reached power. My assessment is therefore surely a rational one. But just how certain does one have to be of forthcoming evil to be allowed commit lesser evils oneself in order to avert it? Who can say for certain, or even in probability, what the assassination of Lumumba averted?

And while it is true that Sendero was defeated by the methods employed, can we know for certain that it could have been defeated only by the methods employed? This would have to be shown for a complete vindication of Fujimori, even on an ends-justify-the-means view of political morality. Again, the answer must be no, we cannot know it. Common sense suggests that a Peruvian catastrophe (if we accept there was going to be one) could not be averted by men with entirely clean hands: but we can't say we know this beyond all doubt, or exactly how minimally unclean those hands had to be, and whether Fujimori kept that uncleanliness to an absolute minimum.

Similarly, we can't know for certain how important Fujimori was to the defeat of Sendero. No man could have defeated the movement single-handed; Fujimori was not engaged in personal jousting with Guzmán in a mediaeval tournament. It was the intelligence services that found Guzmán in hiding and decapitated the movement thereby, the single most important blow ever struck against Sendero; this might have happened whoever was president and had no human infringement of human rights by government forces taken place. Would Sendero have been defeated if Vargas Llosa had been president and had used different methods? Indeed, would Vargas Llosa have used different methods? Again, we can't know.

By contrast, we can know more or less for certain that such-and-such a person was killed illegally, and at such-and-such a person's orders. We know the harm Fujimori did; we don't know the evil he averted, if indeed he really did avert it.

If I had been President of Peru at the time when it looked as if Sendero might win, and that Guzmán might never be found, could I have been persuaded that extra-judicial killings were necessary to defeat it? I hope I am not revealing a disgraceful character when I say that I think I could have been so persuaded. I am not at all sure I should have been able to face down commanders in the field who told me they were necessary, or that my high-minded phrases about the end not justifying the means would not have dried in my throat as I uttered them. This is not to say that I would have been right; I am only relieved that I have never been put in the way of such temptation and that no such responsibility has ever devolved on to me.

10
Crime and Punishment

L ike most social questions, that of the correct response to the phenomenon of crime is unanswerable, if by an answer we mean something that is indubitable, beyond question, and true for all time. Circumstances alter cases, and circumstances are themselves always altering. Nevertheless, we cannot take refuge in the eternal flux of the world: we have to act somehow, even – or always – in the possession of less than complete knowledge.

Recently in my house in France I had two English guests, one who was what might be called a hard-liner with regard to crime, and the other a liberal. By analogy with the Cold War, we might even call them the Hawk and the Dove.

The Dove, of course, was concerned about the causes of crime. These were multiple and complex, not to be fully apprehended by the mind of Man, but nevertheless connected in some way with social injustice. The evident fact of unmerited inequalities in our societies was enough to provoke crime. (Who will deny that, even in a meritocracy, some are born rich, others achieve riches, while others have riches thrust upon them?) On this view, then, crime is an inchoate attempt at restoring perfect justice to the universe.

In favour of the Dove's outlook may be mentioned the equally indisputable fact that most criminals emerge from highly unfavourable circumstances, circumstances that they did little or nothing themselves to create. In my career as a doctor in prison, I did meet a few criminals who were born with the silver spoon in their mouth, and who went to the bad despite their advantages; but their number was trifling by comparison with that of those who experienced deprivation, cruelty, hardship or violence in their childhood. It seems elementary humanity, therefore, to have some sympathy with and for them, and not to victimise them further by condoning punishment. A better approach would be to create social conditions in which there was no childhood deprivation, hardship etc. Punishment is at best a plaster over an unhealing wound, and will never eliminate crime. It is the causes of crime that need to be addressed.

The Hawk would have none of this, of course. Leaving aside the Dove's failure to distinguish between unfairness and injustice (a very large philosophical topic), he pointed out that if it was true that most criminals were deprived in their childhood, it was also true that most people who were deprived in their childhood were not criminals. There is therefore considerable margin for the operation of what is usually called free will. Moreover, if the connection between life history and crime were as described, it could as easily lead to the most illiberal conclusions as to liberal ones.

If it is really true that certain childhood conditions lead inexorably to criminality then, in the absence of any proven technique to break the connexion, this is as much an argument for preventive detention as for leniency. There is, of course, no such technique. Since society must protect itself from criminals, the presence of a deprived background would constitute an argument for longer, not shorter, prison sentences.

The Hawk pointed out, furthermore, that one must not confuse the causes of crime with the appropriate response to criminality once it has developed. And this is so even if one disregards the probability that how society responds to crime is one of the factors a person takes into account when deciding to commit a crime (the decision so to commit being the proximate cause of all crime).

Thus, if as a matter of fact imprisonment prevents the criminal from re-offending, it is quite beside the point that he commits crime in the first place because (shall we say) his mother did not love him enough in childhood. What society is interested in is the prevention of further crime; it cannot engage upon the task of giving him a different past or (slightly less impossible, perhaps) of nullifying the effect of that past.

The Hawk then horrified the Dove further by citing evidence that, contrary to what is often said, prison is actually very effective in the suppression of crime. Indeed, it is the only thing that is effective. For example, offenders sent to prison the first time they are caught (which, of course, is rarely the first time they offend) have a recidivism rate lower than those who receive other kinds of sentence.

Moreover, prison is not a university of crime as is often alleged. If it were, one might expect that prisoners sentenced to longer terms had higher degrees in crime: that is to say, were more likely to re-offend. But in fact they are less likely to do so; prison is therefore the place where criminals learn (eventually, for they are not quick learners on the whole) not to re-offend.

But, said the Dove, if what the Hawk was saying were true (and the Hawk, being a professional writer on the subject had devoted much more time to the study of it than the Dove had done), it would lead naturally to conditions in Europe with regard to imprisonment that resembled those in America – and the Dove would hate that, indeed could think of nothing worse or less acceptable.

This, I need hardly say, was not the end of the discussion. What exactly, asked the Hawk, as so terrible about the American example? Well, said the Dove, they have more than two million prisoners over there. But what is terrible about that, asked the Hawk, if they have all been sent there by due process and are, in fact, criminals (except for those mistakes that are consequent upon any system of criminal justice whatsoever)?

But some races are imprisoned more than others, said the Dove; this hardly seems fair. But, said the Hawk, a differential rate of imprisonment is not in itself evidence of injustice; one would hardly wish to increase the number of Chinese in American prisons simply to bring their proportion up to that in the general population.

The Hawk was a passionate bird, and began to tremble with excitement (I know the symptoms well, and try, somewhat unsuccessfully, to control them in myself). He pointed out that it is completely absurd to dwell on the prison population as a proportion of the general population. To have but one prisoner in a country in which there had never been a crime would be an outrage. What counted was not the prison population as a proportion of the general population, therefore, but the prison population in relation to the number of crimes committed.

Now if Britain, which has gone in half a century from being a country with a low crime rate to one with among the highest rates of crime in the western world, had the same sentencing policy as Spain – that is to say, if it sent people to prison for the same reasons and for the same length of time as in Spain – its prison population would be not 80,000 but 400,000. Not coincidentally, Spain is a country whose crime rate is – yes, about one fifth of Britain's. Furthermore, said the Hawk, if Britain had 400,000 prisoners, it would have the same proportion of the population in prison as – yes, the United States.

Furthermore, it has been estimated that if Britain now had the same sentencing policies as it had in Edwardian times, its prison population would be – well, about 400,000. According to the Hawk, the crime rate in Britain started its vertiginous rise after, and not before, the sentencing policy became weaker, as a result of years of Dove-ish propagandizing; I did not know enough either to agree or to disagree with his historical analysis, but I (who was much more in sympathy with the Hawk than the Dove) added my mite, to the effect that to fail properly to punish and disable criminals from committing further crimes was a failure to protect the poor, given two cardinal facts: first, that if it is true that the vast majority of criminals are poor, so it is also true that the vast majority of their victims are also poor; second, that the class of victim is always very much larger than the class of perpetrator.

Perhaps it will come as no surprise to learn that no minds were changed in the course of this argument: after all, one argues for victory, not for truth. However, I suspect that the Dove might be slightly less dove-ish in the future, should the argument recur in other circumstances and surroundings, with other people, without (for temperamental reasons) undergoing a full avian metamorphosis. For those with a soft heart, the problem with the Hawk's argument is this: that while long imprisonment causes tangible distress to certain easily-imagined individuals, the harm therefore appearing concrete, the people to whom good is done by the use of imprisonment because they are prevented from becoming victims of crime remain shadowy, and therefore the good is purely abstract or notional. It is for this reason that Hawks always have a public relations problem.

Careful readers will no doubt have noted the use by the Dove of the United States as a kind of bogey-man whose example is at all costs to be avoided. This, of course, is the polar opposite of the United States as the last great hope of mankind. In Europe, the United States is often used as a trope for all that is bad about the modern world, and in particular as an example of a savage, unsocial world, a kind of Wild West of the soul, where everyone is selfish, concerned only for his own advantage, indifferent to the fate of others, crassly materialistic, and so on and so forth. Although no European visitor ever claims to have seen such a thing for himself, many Europeans conceive of the United States as a land in which, if a person is injured or falls ill in the street, he is left to die there if he is not privately insured.

I am not myself an idolater of the United States. I do not believe that all that is American is best. It is neither a model to be imitated in all things, nor a model at all costs to be avoided. Its manifest failings are exceeded by its manifest virtues: but it requires discrimination to decide what is worthy of emulation and what of avoidance. Generally speaking, we in Europe get things exactly the wrong way round.

For me, the high imprisonment rate in the United States is a sign of social health, not of social disease. Of course, I do not approve of any miscarriages of justice or of incidents of brutality that occur in American prisons: but when I compare the confidence and resolution with which America faces the problem of criminality with the vacillation in most of Europe (some countries excepted), I cannot help but be struck by the difference, which is all to our disadvantage. The American system, for all its faults, is prepared to draw a line; European systems, on the whole, are not. But my view is exactly the opposite of what most Europeans, or at any rate educated Europeans, and no doubt many Americans, think.

I am aware that those who in the past advocated the stern repression of crime have a bad record for cruelty. However, the conditions of the past are not those of today, and we have to deal with our own problems as best we can, without either blindly following or rejecting the past. In other words, we have to think for ourselves. I do not believe that absolutely anything is justified in order to reduce crime, and I believe (contrary to many Hawks) that prison should not be a kind of Golgotha for the prisoners. But societies such as several western European ones that cannot summon the confidence to set apart those who have persistently shown themselves unwilling to abide by the most elementary rules, and which prevaricate and beat their breast wondering how they and not the law-breakers are really to blame, may truly be described as decadent.

11
The Cult of Insincerity

I once had a patient who had had the words 'Fuck off' tattooed on his forehead in mirror writing. When I asked him for the reason for this, he said that it was to wake him up in the morning when he looked at himself in the glass. It never failed, he said.

Newspapers perform more or less the same function for me. There is always something in them to irritate me profoundly, and there is nothing quite like irritation to get the juices circulating and the mind working. Oddly enough, only the print version of a newspaper, not the online one, has this tonic effect upon me; perhaps this is a conditioned response. I am like one of Pavlov's dogs, who salivated at the sound of a bell. I have only to hold a newspaper in my hand to feel a pleasant frisson of outrage coming on.

Whenever I am in France, I read the French newspapers (the French read fewer newspapers than any other nation in the western world, by the way). There is always plenty in them to infuriate me, and so they are well worth the reading; for it must be confessed that indignation is one of the most rewarding of all emotions, as well as one that automatically gives meaning to life. When one is indignant, one does not wonder what life is for or about, the immensity of the universe does not trouble one, and the profound and unanswerable questions of the metaphysics of morals are held temporarily in abeyance.

The other day – well, on Saturday, 5th September, to be exact – I opened *Le Monde* to the page called 'Debates.' The page was devoted to prisons in France, where conditions are acknowledged by almost everyone to be very bad. The prisons are overcrowded; there is much violence between prisoners; the staff, according to Dr. Dominique Vasseur, who wrote a best-selling book about her time as a doctor working in the largest prison in Paris, are callous and often corrupt. If her book is to be trusted – and no one, I think, has suggested that she was lying or grossly exaggerating – prisons in France are far worse than those across the Channel, which themselves are by no means always model institutions.

Prison reform is an honourable cause; and while I don't agree with Churchill, that a nation's level of civilisation can be gauged by the way in which it treats its prisoners, I have always opposed the brutality that can so easily pervade what Erving Goffman called 'a total institution.' In the prison in which I worked, I insisted to the staff that their ascendancy over the prisoners must be moral rather than merely physical; and that, while they could be sometimes stern, they must always be fair. Moreover, they should always remember that, in prison, small things become large; and therefore, if they have promised something to a prisoner, they must always fulfil their promise. For otherwise the prisoner will be eaten up by a sense of grievance, and there is nothing like grievance to prevent a man from examining his own responsibility for his situation.

But half the page of *Le Monde* was taken up with a plea for the greatest reform of prison of all: total abolition. It was written by a teacher of philosophy at a *lycée*, one of the elite state schools of the country; and if it were not for the fact that many young people tend to believe exactly the opposite of what their teacher teaches them, I would have said that he must be a corrupter of youth. It is odd that a man who presumably has spent a large part of his life on abstract questions should show such little capacity for critical thought. In him, at any rate, the Cartesian spirit is dead.

The article's title is: "An absurd system in a modern democracy." The headline continues: "Over and above humiliation, it has become more murderous than the death penalty."

The evidence for the latter assertion is that, since the abolition of the death penalty in France *de facto* in 1977 and *de jure* in 1981 (incidentally, you'd think it was BC 1981, the way Europeans look down on countries like India and Japan that retain the death penalty), at least 3000 prisoners have committed suicide in prison. And this fact alone is taken by him as indicating that prisons should no longer exist in France.

He writes:

The abolition of the death penalty brought about by the Left appeared logically and sociologically unavoidable; but it was only paralogical and paradoxical. It must be recognised: suicide kills more in prison than the death penalty ever did.

Only the abolition of prison, of course, will prevent suicide in prison. I leave aside the question of what 'unavoidable' means.

The malign influence of Foucault is everywhere in the article. Foucault has demonstrated that the end of cruel public punishments consecrated the arrival of the modern state which manifested its power hidden from view.

Personally sado-masochistic, Foucault tried (using an entirely bogus historiography) to demonstrate that humanitarian reform was actually nothing of the kind, but the replacement of one kind of raw power by another, more hidden and therefore dangerous and sadistic power.

Using precisely Foucault's paradoxical thinking, the author writes:

The abolition of the death penalty therefore constituted less the symbolic accession of the left than the event that signified the defeat of its thought. Far from resolving a moral and political problem under the banner of the rights of Man, the abolition of the death penalty in 1981 sanctioned and sanctified punishment as incarceration. The left ratified a vast tendency in society in which squeamishness vies with hypocrisy.

The argument seems to be this: that the abolition of the death penalty led to an increased number of prisoners, which in turn led to an increased number of suicides among prisoners. Therefore the abolition of the death penalty was not a humanitarian measure.

I will not comment on the empirical evidence, or lack of it, for the assertion that the number of prisoners increased because of the abolition of the death penalty. Nor will I ask whether the increase in suicides after the abolition of the death penalty was predictable, as it would have had to have been if the abolition is to be designated as hypocritical. (We can blame people for not knowing that there are unpredictable consequences of their acts, but not for not knowing what those unpredictable consequences are.) Nor will I point out that there are rather obvious moral differences between an execution carried out by the state and a suicide, even that of a prisoner in the state's care.

Let us, for the sake of argument, accept what the author claims. It would seem to entail the rather odd conclusion that a restoration of the death penalty would be a humanitarian measure. It would reduce the total number of deaths by reducing the prison population, and therefore the number of suicides in prison (assuming what is highly probable, that the rate of suicide among prisoners is higher than it would have been if they had not been imprisoned). On this view, the death penalty is a kind of expiatory sacrifice made on behalf of the whole population, rather than just a punishment properly so-called.

This, of course, is not the conclusion that the author draws. Rather, he wants the abolition of prison. As we shall see, he even wants something even more radical that that.

One thing that is notably absent from the article is any notion of crime or of the effects it has both on individual victims and on society as a whole – in the sense that a lot of crime causes fear and alters the mentality and behaviour of almost everyone in the direction of mistrust, caution and loss of freedom. It is as if only the criminal, and neither his act nor his victim, were of any interest to the author.

He suggests liberating prisoners 'who can leave prison only humiliated, raped or desperate.' For him, prison should be nothing but a therapeutic institution, one that does the prisoner good; if it fails to do that, it fails to do anything.

Again, I will pass over the question of whether humiliation is always and everywhere a bad thing. After all, the prospect of humiliation is one of the things that keeps us upright, as a cane keeps many a rosebush upright. We are social beings because we have a capacity to feel humiliated – or it might be the other way round. Be this as it may, there could be no prospect of humiliation if there were no actual means by which we might be humiliated. I am not particularly criminally-inclined, no more than average I would say, but I have often been kept on the straight and narrow path that leads to respectability by fear of humiliation. However, let us leave aside the interesting question of the necessary dose of humiliation necessary for the maintenance of society.

What would our author have instead of prisons? He says that he would build institutions designed by men and women who really wanted to look after wrongdoers, not institutions built by '*betonneurs*,' those who construct in raw concrete. (Here, in his contempt of those who build in concrete, I agree with him.) But what kind of institutions would these be?

Here we come to the heart of his outlook, and that of many like him. He says that those prisons that are salubrious as buildings should be converted into 'places of social reintegration,' not only for those who have committed a crime, but for those 'socially disintegrated people' who have committed no crime: tramps, perhaps, or schizophrenics in need of rehabilitation. In other words, criminals are not to be marked out from any other people with difficulties of one sort or another, or treated differently from them.

The desire to blur limits and boundaries, in order to overturn society, has long marked out a certain kind of leftist. Because in social phenomena there are always borderline cases, they wish to undermine the very idea of categories. They are like people who would deny that anyone is tall because there is a fine gradation between tallest and shortest. Thus, because some things were considered crimes that are so considered no longer, and some things that were once legal that are now deemed criminal, they deny that the crime is anything other that an arbitrary social construction. A criminal is someone who merely has difficulty in his relations with society as some men have difficulties in their relations with their wives (and vice versa). What more natural, therefore, than that they should all attend the same day care centre, where they will be cured of their difficulties by psychological means?

'It is necessary,' says the author, 'that the punished person should understand his mistake.' Prison is obviously not the place for this; he comes out with as little understanding of his 'mistake' as he went in with. He therefore needs some kind of psychotherapy until he gains the requisite insight. We can see the Socratic paradox underlying this: that no man does wrong knowingly. There is no such thing as a wicked man.

This does violence not only to our knowledge of the behaviour of others, but to our self-knowledge. Which of us has never done wrong knowingly? Indeed, under most jurisdictions, a person is not guilty of a crime unless he has the requisite *mens rea*, a guilty mind, which implies the ability to have acted differently if he had so chosen.

There is no recognition whatsoever in the article that the purpose of criminal law is to protect the population from criminals, not to make criminals better people. Of course, it would be nice if they became better people, as indeed they often do with the passage of time; but criminal justice is not group therapy. It is, moreover, preposterous, and deeply condescending, to suggest that criminals do not know what they are doing, and that what they need is therefore some kind of help to know it. As for calling crimes a 'mistake,' equivalent, shall we say, to putting the wrong postage on a letter or forgetting to put salt in the soup, it empties the world of all moral meaning whatever.

There is in the article a moral exhibitionism, which is generosity of spirit at other people's expense. This, I think, is one of the sicknesses of our age, the desire to appear more-compassionate-than-thou. I suspect that, in his heart of hearts, the author does not believe a word of what he says: a thing common among intellectuals.

12
Let Them Inherit Debt

R ecently I was in public discussion on the question of poverty in Britain and how to overcome it. The poverty is relative, of course, not real destitution; but there is nonetheless no doubt about its squalor.

Among the panel on which I appeared was a very pleasant member of the Fabian Society, that is to say the society whose goal was to achieve socialism in Britain gradually and by reformist means (it was named after the Roman general, Quintus Fabius Maximus, who wore Hannibal down by attrition rather than by pitched battle).

One of the definitions of poverty most commonly used (and which smuggles in the supposed inherent desirability of equality) is that of an income less that 60 per cent of the median income. In a society of billionaires, therefore, a millionaire could be said to be living in poverty. And it is clear on this definition that a society could be become much poorer as a whole, and yet have less poverty than it had before when it was richer.

But let this pass. One of the Fabian's suggestions to bring about a more equal society and thereby lessen poverty was to increase and extend inheritance tax. The money raised would be distributed in one way or another to the poor (minus deductions, of course, for the pay, perquisites and pensions of those who had to administer it, a proportion not likely to be small). For, as he said, it was unfair that some people, by accident of birth, should inherit wealth while others should inherit nothing.

It seemed to me obvious that, underlying and if you like impelling the proposal, was our old and trusted psychological friend, the one who never lets you down, resentment. Why should some people, no better than I and sometimes much worse than I, be better off than I, merely by chance, that is to say by accident of birth? Why should some people be handed on a plate what I have to work all my life for, or indeed in some cases more than I can ever hope to earn and accumulate?

Nothing could be less fair.

It is unfair, but is it unjust? The argument certainly has the appeal of plausibility to all those whose station in life is not what they would like (which is quite possibly the majority of mankind).

There are many unfairnesses in life that we must learn to put up with, if we are to have any chance of happiness or even of tolerable contentment. For example, I should like to be taller, better-looking and more intelligent and gifted than I am. Every time I meet someone better-looking than I, taller than I, or more talented than I, which I do very regularly, I experience a brief spark of envy. What did they do to be as they are, my superiors? Why did providence, or chance, endow them with characteristics so much more attractive than my own? Needless to say, I never stop to think that, just possibly, some people might ask the same of me when they meet me.

But the differential endowments of nature are unfair, not unjust, because (at least as yet) no human intervention can prevent them. The inheritance of wealth is not like this: it is a human arrangement that could be abrogated if not easily, for political reasons, at least with some effort. And if injustice is unfairness brought about by human means, then inheritance of wealth is unjust. Ergo, inheritance of wealth ought to be forbidden because it is unjust, and we must always seek justice.

The question, then, is whether we should always seek justice to the exclusion of other desiderata. Is it true that justice always and everywhere trumps other considerations? I think the answer is no.

Let us widen the question of what we inherit. Although I have personally inherited very little in the way of money or material goods, I happen, quite by chance, to have been born into a comparatively rich country, relative to most of the world's population. This has given me the opportunity to live at a much higher material standard of living than most of mankind born elsewhere (though the unwisdom of our government and the improvidence of its population is fast eroding these advantages). Am I required, ethically, to renounce the fruits of the advantages that I have inherited quite by chance, and live at the median standard of the whole of mankind? Or do ethical considerations of this kind stop at constituted borders – which would be indeed a strange thing?

Let us go a little further. By having been born where I was, I had a life expectancy that much exceeded that of the great majority of mankind throughout its history, and (to a lesser extent, of course) even that of my own parents. This unearned benefit accrued to me again by chance; and, if I wanted to feed my resentment, I could work myself into a fury by reflection that generations subsequent to mine, that from all appearances are even worse than mine in many respects, will live longer than I, again by virtue of having, by chance, been born after me. Could anything be less fair?

Now it is quite clear that the progressive increase in life expectancy is not a work of nature: it is the work of man. The greatest contribution to the increased life expectancy has been brought about by the improvement in living conditions, itself the result of technical and to a much smaller extent political progress; all of it the work of man. In addition, whole diseases have now been eliminated by medical progress. I myself, had I been born in the eighteenth century, would long ago have died in a state of hopeless dementia, easily and routinely prevented in the twentieth century by the simple expedient of taking two small tablets a day. But I contributed nothing to the scientific research that made this possible, which took place before my birth; I have inherited a benefit greater, in my case, than any amount of money could ever have been.

One could go on and on giving illustrations of a like nature: indeed, the illustrations, if they were to be exhaustive, would stretch out nearly to infinity. Even when I do something as banal and everyday as switching on a light, I receive an inherited benefit unknown to men throughout the great majority of human history.

In short, civilisation itself requires the gradual accretion of unearned benefits. No matter how great or intelligent or gifted a man may be, he builds upon what he has inherited; he is never entirely self-made, as if he started out life in a little personal Garden of Eden, and had himself to eat of the fruit of the Tree of Knowledge. One of the mightiest minds in the history of mankind, that of Sir Isaac Newton, put it succinctly: If I have seen further, it is only by standing on the shoulders of giants. Mozart himself, another astonishing genius, did not start from nothing: he inherited a tradition, and worked on it. Maybe, as is sometimes said, he took dictation from God; but to do so he had to learn how to take dictation.

But, the Fabian might reply, civilisation is the inheritance of all mankind, and therefore there is no unfairness in its inheritance. Alas, this will not do, because it is an unavoidable characteristic of civilisation that it is inherited unequally, even within a single society. One child may be born into a family of the highest cultivation and intellectual attainment, another, of equal ability, into a family of swinish degradation. The latter may reach high levels of civilisation, but if he does it will be after incomparably greater labour than the former had to employ to reach them.

Only the most radical and horrifying social engineering could eliminate these differences. Indeed, a world in which there were no such differences would have to resemble very strongly that depicted in Aldous Huxley's *Brave New World*. Such a world would be perfectly just, in the sense that it removed all unfairness brought about by human action and activity, but it would also be perfectly horrifying.

The desire to give one's children an easier and better path through life than one has had oneself is also not to be deprecated. Of course, in a sense the attempt is based on an illusion: a better path through life is not a matter merely, or even principally, of material wealth. Nevertheless, few of us would prefer to live among ugliness than beauty, in squalor than cleanliness, with bad food than good, with difficult access to culture than easy, and so forth. And, while we might want all these things for the whole of mankind, we want them more for our own children than for the children of others. A world in which no parent was especially concerned with the well-being of his own children would be a truly dehumanised world, a world in which every mother would be a Mrs. Jellyby, who neglected her own children while concerning herself with the natives of Borrioboola-Gha (on the left bank of the Niger).

> The African project [she said] employs my whole time. It involves me in correspondence with public bodies and private individuals anxious for the welfare of their species. It involves the devotion of all my energies, such as they are; but that is nothing so long as it succeeds; and I am more confident of success every day.

Meanwhile, her hair remains unbrushed because 'she was too much occupied with her African duties to brush it;' she does not notice that her own children fall down the stairs because 'her eyes had a curious habit of seeming to look a long way off, as if they could see nothing nearer than Africa.'

What Dickens portrays in his deeply and irresistibly comic fashion would actually be horrible if put into practice on a universal, or even on a large, scale, but that is what the project exemplified by the wish to suppress inheritance as unjust would logically entail. However, it has not pleased our creator – whether God or the evolutionary process – to give us the capacity to feel for the whole of humanity equally. Our emotional responses, our genuine concerns, are not indefinitely expandable; and the requirements of perfect justice make us not more, but less, human.

As usual, Shakespeare has something to say on this, at least by implication. When the players come to Elsinore, Hamlet asks Polonius to treat them well. Polonius replies, 'My lord, I will use them according to their desert', to which Hamlet replies, 'God's bodkin, man, much better. Use every man after his desert, and who shall 'scape whipping?'

In other words, assuming justice to be the allocation of result by desert alone, we should all be in a pretty pickle if there were justice in the world, a fate from which (in fact) only the inheritance of mercifully unjust privileges can save us.

Behind the proposal to abolish inheritance – for if one takes the proposal seriously, why should anyone have the right to inherit anything? – lies a desire for the *tabula rasa*, the blank slate, on which, to quote Mao Tse-Tung's famous, or infamous, words, the most beautiful characters can be drawn. To start again, to start anew, to make the world a new Garden of Eden, is the goal.

But why is that goal so attractive to so many, at least in the modern world, and particularly among intellectuals? I think the answer is egotism and self-importance. Having lost his religious faith in a being much greater than himself, modern man finds the existential limitations imposed upon him by nature to be meaningless, arbitrary and offensive.

Not Man, but each individual man, becomes the measure of all things. He cannot stand anything that he has not himself fashioned, and so the world must be made anew, over and over again, with no generation ever admitting its debt to the previous one, or thinking seriously about the succeeding one.

The strange thing about the proposal to abolish inheritance is that it usually goes along with an attachment to deficit financing: spend now, pay later, and for many years, or even centuries, to come. We may not leave our children our houses, but it is perfectly all right to leave them our debts, a curious morality to say the least.

13
We Are All Guilty

I was in New York when Lehmann Brothers collapsed and I was in Dubai when property prices fell there by fifty per cent in a week. I claim for my presence no causative relationship to these unhappy events, of course, but it did occur to me that I could start an investors' newsletter, and charge for it, that consisted solely of my travel plans.

While in Dubai I bought and read a book about Bernard Madoff: the subject of pyramid schemes seeming only too appropriate to the time and place. Alas, Mr Madoff is a boring villain, not at all the swashbuckling type that one likes one's swindlers to be, at least in literature or when swindling someone else. I remember when he was arrested that he looked just the kind of solid, respectable, level-headed yet intelligent person to whom I (who take no interest in my own financial affairs) would like to entrust my money, such as it is.

The book was a disappointment, really: not only did Mr Madoff not appear to have any bad habits of an interesting kind, apart from malversation of funds, but at the end of 320 pages I still did not know when he started his scheme, how much exactly he had stolen and who was in it with him. The financiers around him tentatively suggested to the author different dates for the commencement of his nefarious activities, though perhaps some of them were feigning uncertainty only in order not to implicate themselves.

Mr Madoff's originality seems to have consisted largely of offering not fabulous, but steady profits; his pretence of being indifferent whether anyone invested with him or not; and the successful creation of an impression that his fund was for an elite, not for the *hoi polloi*. One of the reasons for his success (if success is quite the word I seek) was his appeal to snobbery. His story is proof, if proof were needed, that in finance, as in art and science, originality is not in itself a virtue.

Like many people, no doubt, I have been reflecting of late on an economic crisis that does not yet appear to me to be quite over, since many of its causes are still in operation, and despite the recovery everywhere of the stock markets. The crisis hardly affected me personally, but no man is an island and all that; besides, crises have a habit of eventually engulfing even those who thought themselves immune from their effects. It would only take a little inflation for me to start feeling some serious anxiety on my own behalf.

In the meantime, the search for people to blame continues. This is hardly surprising, because blaming is so much more fun than the supposedly more fruitful and intellectually mature activity of explaining. However, in human affairs the two activities are often very similar or at least difficult to disentangle: explanation and blame are often like King Hamlet's funeral and Queen Gertrude's second wedding:

Horatio: My lord, I came to see your father's funeral.

Hamlet: I prithee do not mock me, fellow-student.
I think it was to see my mother's wedding.

Horatio: Indeed, my lord, it followed hard upon.

If widespread fraud, greed, stupidity, wilful ignorance and gross improvidence do not occasion censure, it is rather difficult to see what would. And if nothing occasioned censure, the world would be emptied of moral meaning.

There can be no doubt, I think, that bankers and assorted financiers have been amply excoriated, as far as I can see quite rightly, in the press and elsewhere. Nor have governments entirely escaped their share of the blame. Both by acts and omissions they created the circumstances in which crooked, or at least less than scrupulous, practices could escape detection. Moreover, all western governments (as far as I can see) have been operating Madoff schemes for many years, the main difference between Mr Madoff and themselves, apart from the matter of scale, being that governments can coerce contributions while Mr Madoff could merely solicit them. Whether this makes western governments or Mr Madoff the more villainous, I leave to moral philosophers to decide.

Of course, there are those like the late J K Galbraith who argued that government deficits were a good thing, a sign almost of economic health and well-being. He sometimes, indeed, made it sound as if the larger the deficit, the better for all concerned, and the healthier the economy. As Galbraith pointed out, the infrastructure built today on borrowed money benefits future generations, so it is only just that they should be left with some interest to pay on the loans that made possible the things from which they derived such benefit.

It is obvious that not all borrowing is economically burdensome, and indeed credit is the life-blood of large-scale economic activity. But it is an elementary error to suppose that if the existence of 'a' requires the existence of 'b', then the existence of 'b' is evidence of the existence of 'a'. It is one thing to borrow money to start up a company that will soon grow and be profitable, and another to borrow the same sum to go on the holiday of a lifetime, whatever appalling thing that might be. It isn't even true that all investment is wise or profitable (communist countries long had very high rates of investment), let alone all borrowing. And the fact is that we have borrowed enormously that we might live well today, and not that we, let alone our descendents, might live better tomorrow. Future generations, I suspect, will have good cause to curse us when we are safely in our graves.

It is, of course, possible that future economic growth will render current borrowings nugatory or trivial: all one can say, however, is that in the race between growth and borrowing, the latter seems to be winning hands down. Debauching the currency might be a solution to the problem in a certain sense, but it undoubtedly has what these days people call a downside.

Having blamed the bankers and the governments, however, it is now time to turn the spotlight of blame onto ourselves: if by ourselves we mean the common or ordinary people. Here, on the whole, criticism has been much more muted or reticent, no doubt because it is not exactly *bon ton* in our democratic age to suggest that ordinary people are fully as capable of every human vice as those who rule them.

But the fact is that it is not only governments that have been improvident and have spent well beyond their means; millions, scores of millions, of perfectly ordinary people have done so as well, and have behaved not only as if there were no tomorrow but as if there could be no tomorrow. In so far as they thought about their debts at all, they thought they could merely walk away from them, as if to do so were of no moral or characterological significance. Another day, another default, seems to have been their motto.

It could, I suppose, be said in their defence that almost everything possible has been done to encourage them in their improvidence. In the United States successive governments encouraged, indeed required, banks to lend money to people whom the banks knew to be bad risks; in Britain, a government stood by while banks offered mortgages of 125 per cent to people in areas where unemployment was not exactly unknown, simply because its own popularity depended upon the illusion of prosperity created by the asset inflation which easy money provoked. It was not difficult, either, to find in our newspapers statements by supposedly serious journalists and commentators to the effect that we had at last entered an age of the uninterrupted virtuous circle, or rather spiral, in which growth created rising demand which called forth yet more growth which created... etc. etc. The business cycle was abolished; there would be no more downturns to embarrass those who had overextended themselves by borrowing what to them were vast sums.

No doubt it was very wrong of banks to offer credit to the uncreditworthy: but while you can lead a man to a loan, you cannot make him borrow. To give a feckless man a credit card is both wrong and feckless; but the man to whom it is given does not cease thereby to be feckless himself when he spends money he is never going to have.

Moreover, the figures for personal indebtedness, in America, Britain and elsewhere, suggest that fecklessness or improvidence are far from being the characteristics of a few individuals but have rather become almost normal, at least in the statistical sense. The idea of cutting one's coat according to one's cloth, or of taking pride in owing nothing (in the financial sense) to anyone, has disappeared among us.

The decline, if not the total absence, of personal providence in the general population will no doubt be the subject matter of doctoral theses of future history students – if, that is, there are any.

Let us imagine the academic controversy occasioned two centuries hence by the appearance of the book (in whatever form books then appear) entitled *The Decline of Personal Financial Probity and Prudence in the West, 1950– 2010*. The disputants will divide into two main camps, the materialists and the idealists.

The former will ascribe the decline to economic conditions rather than to anything that happened in the minds of those among whom the decline was evident. For example, in times of inflation (whether of ordinary goods or of assets) what counts as providence changes fundamentally. A squirrel that hides nuts for the forthcoming winter is not in fact provident if the whole forest is about to be cut down and flattened, and the earth churned over. The virtues of a previous age do not answer to the needs of a new age; providence in an age of asset inflation is, in fact, its opposite, improvidence. As Marx said, it is not consciousness that determines man's being, but on the contrary, his being that determines his consciousness.

But, reply the idealists, what happens in the aggregate is only a summation of millions or hundreds of millions of individual decisions.

How people behave is determined by what they believe; and if they come to believe that slow accretion is the policy of fools and that maximum consumption in the here and now is the only meaning of life, it is hardly surprising if what you get is an orgy of speculation combined with insouciant expenditure.

Of course, there is a wealth of question-begging in the phrase 'if they come to believe,' for how do people come to believe anything? Why do they change their minds, such that those things that once seemed to them good now seem to them bad, and vice versa? Still, enquiry must stop somewhere if we are to hold an opinion about anything; the search for ultimate or final causes of social phenomena often conceals a cowardly refusal to say anything that could possibly be contradicted by anyone or that risks refutation. And it doesn't really matter how people come to believe what they do believe, so long as it is accepted that what they believe is what causes them to behave as they do.

No doubt a twenty-third century revisionist historian will then come along and say that the whole debate is beside the point in any case, since there was no decline in the financial probity and prudence of the population in the years specified. This will be proved by a re-working of the statistics, which will show that the supposedly high levels of personal indebtedness were really nothing of the kind. Every layman will end up thoroughly confused and not knowing what to believe.

Really, though, it is all quite simple. Our banks were no good; our government was no good; and we were no good. Apart from that, everything was fine.

14
Please Feel My Pain

S hortly before Mr Blair was elected Prime Minister of Great Britain, a newspaper discovered that I had not had a television in my home for about thirty years. This struck the editor of the newspaper as an extraordinary circumstance, as extraordinary as if I had been an anchorite in the Syrian desert subsisting on locusts and honey, and he contacted me to ask me if I would agree to having a television installed in my home so that I could tell readers, after a week of watching it, what I thought of it. This I consented to do on one very firm condition: that the newspaper took the television away at the end of the week. The newspaper agreed.

When the television arrived, I plugged it in and turned it on. The picture was grainy, for something else was required, evidently, to have a good reception. But it was good enough to know what was going on.

The programme was one of those in which a degraded family, or perhaps I should say a group of human beings who have lived in close association for some time or other, airs its appalling behaviour in public in return, I should imagine, for money, and for the prurient delectation of a voyeuristic audience.

A fattish woman approaching middle age was complaining in a monotonously high-pitched voice, halfway between a harangue and a wail, about her three daughters who were aged twelve, thirteen and fourteen respectively. According to her, they 'did drugs' and had left home to be prostitutes.

At this point, the presenter of the show interrupted her and asked the audience to give a warm welcome – with, of course, a round of applause - to the three young trollops in question, who came tripping down the steps to the television set with smirks of self-satisfaction on their faces. No lack of self-esteem there, I thought; rather too much, in fact.

Of course, mother and daughters began at once to trade high-pitched insults and accusations, and generally behaved like a dog and a cat enclosed in a sack. There was undoubtedly a morbid fascination in all this, though the spectacle was disgusting; suffice it to say that I was not encouraged by it to take steps to ensure that the television had a permanent presence in my home.

The newspaper had given me a timetable of programmes to watch, though it did not inform me as to the criterion it had used in their selection. Whether what my wife – who likewise had had no exposure to television for years before I met her – and I watched was better or worse than the average that was on offer to viewers, we could not say; but it seemed terrible pabulum to us, having approximately the same effect on our consciousness as a food-mixer on vegetables. It turned it into a kind of soup.

One of the programmes we were enjoined to watch was a breakfast time chat-show. The chief guest on this occasion was a man called Tony Blair, who at this stage was only Leader of the Opposition, though it was clear by then that he might very well be the next Prime Minister.

However, both my wife and I thought that the man on the television was not really Mr Blair, but someone imitating him, a clever impersonator perhaps, satirising the state of British politics. We had never seen or heard Mr Blair before, of course, or any other contemporary politician (one of the inestimable benefits of not having a television, and of never listening to the radio), so we had no standard by which to compare the guest of the breakfast TV with the 'real' Mr Blair. We both of us independently assumed that the preening, self-satisfied young popinjay could not possibly be the real thing.

Alas, we were wrong. In so far as anything about Mr Blair could be called real, this was the real Mr Blair. In the years that followed, the fact emerged - though it took a long time for many in the population to understand its significance - that thinking about Mr Blair's mind and its verbal productions was a little like looking at the drawings of M C Escher, the Dutch artist, 'in which lines of people ascend and descend stairs in an infinite loop, on a construction which is impossible to build and possible to draw only by taking advantage of quirks of perception and perspective.' (I have quoted from the excellent characterisation to be found on Wikipedia of one of Escher's most famous pictures.)

To look at Escher's work for too long is to drive oneself to complete distraction; and so it is with the contemplation of minds and characters like that of Mr Blair. The problem with an Escher picture is that it undoubtedly exists in itself, so it a tangible object; likewise the minds and characters of Mr Blair and his ilk (the ilk to which I shall return), for they are tangible too in the sense of undoubtedly being existent. But looking at an Escher picture is also to enter a clearly impossible world without being able to put one's mind's finger, as it were, on what exactly it is that makes it impossible, and keeping it there. One oscillates constantly between a sense of reality and its opposite, without ever one or the other achieving final victory. The cleverness of Escher is to make us know and not know something at the same time, which is both infuriating and anxiety-provoking.

We both know and don't know that people like Blair and his model, Mr Clinton, are liars: we don't know it because, while they constantly say things that are demonstrably and patently untrue, they subsequently claim not to have lied because, at the time they said those things, they believed them to be true, with all their hearts and with all their soul: a claim that is very difficult to disprove. And they themselves know that they believed them to be true with all their hearts and with all their soul because they said them, and they never say anything that they know to be untrue, because they are not liars. This is a labyrinth from which, once entered, there is no return.

It would be comforting to imagine that politicians like Blair, Clinton and, I fear, Obama have been visited upon us like aliens arriving from outer space (George W Bush was dreadful in a rather different, and perhaps slightly more traditional, way). But they have not: they were, after all, elected - Blair and Clinton re-elected - by millions and millions of votes. Such politicians are the natural product or culmination of a certain cultural development over the past sixty years; they not only rule us, they represent us and what we ourselves have become.

The cultural development in question is the systematic over-estimation of the importance not so much of emotion, as of the expression of emotion – one's own emotion, that is. The manner with which something is said has come to be more important than what is said. Saying nothing, but with sufficient emotional vehemence or appearance of sincerity, has become the mark of the serious man. Our politicians are, in effect, psychobabblers because we are psychobabblers; not the medium, but the emotion, is the message.

Often in the street I hear people say to one another 'I really, really think that...' But is 'to really, really, really think that...' to provide even stronger evidence for the proposition that follows than 'to really, really think that...', which in turn is to provide stronger evidence than 'to really think that...?' And in what way, precisely, is 'to really, really think that' different from simply 'to think that?'

And how can one contradict someone who really, really thinks something?

Is that not an attack on his or her whole personality?

Of course, the emotion with which an utterance is made has always been important in our assessment of how much to trust or believe the utterer of it. But whereas emotional expression was once the servant of meaning, now it is the other way round: meaning is the servant of emotional expression. Thus, Mr Blair has repeatedly defended himself against criticism by stating that he always did what he thought was best, even when what he apparently thought was best was, to many others, transparently corrupt or self-interested. Not his policy, but his goodness, is being defended.

This tactic worked for a long time. This is because we ourselves often seek indulgence for our own wrongdoing that allegedly derives from a pure heart. We want people to concentrate on the purity of the heart, and forget the wrongdoing.

One sign of the elevation of feeling over thought and reason is the way in which the proof of allegations such as bullying and racism has been emptied of any requirement of objective evidence that they have taken place. In the British public service, for example, a great deal of time is wasted (that is, wasted from the point of view of the ostensible end of that service) in investigating complaints and grievances by people who claim to have been the victim of one or other of these things.

In many organisations, bullying or racism is said to have taken place if the victim, or alleged victim, feels that he has been bullied: and since, of course, the complainant is the final arbiter on the question as to whether or not he feels bullied or the victim of racism, human relations become extremely difficult. Anything at all, any small jest or expression of displeasure or disdain, can be interpreted as bullying or racism. The best form of defence is attack; the alleged perpetrator can lodge a counter-claim against the complainant on similar grounds.

Everyone therefore walks on eggshells all the time, wondering who will be the next to take offence at what; and everyone becomes supersensitive himself to perceived slights and insults. Fragility becomes general, and everyone is on tenterhooks either as a potential victim or potential perpetrator.

As if this were not bad enough, financial incentives are often offered to those who feel victimised: a fact that gives superior levels in the administration immense power over their underlings, for it is the superior levels that must, in the first instance, arbitrate between the complainant and the complained against. But the superior levels can never feel safe themselves, for they operate in the very same atmosphere of fear. The principle applies that was first enunciated with such clarity by Jonathan Swift:

So, naturalists observe, a flea
Hath smaller fleas that on him prey;
And these have smaller still to bite 'em;
And so proceed ad infinitum.

An official report in Britain into the murder of a young black man by five white thugs, which the police through their habitual incompetence (or, quite possibly, corruption) failed to solve in the sense of securing convictions against the murderers, suggested that the legal definition of a racial incident henceforth be any incident that is perceived by anyone, either involved in it or as a bystander to it, as being racial. In other words, there should be no requirement of actual racist behaviour, no objective correlative whatsoever.

In the circumstances it is hardly surprising that so many of us have become like the princess in the story by Hans Christian Andersen, who could not sleep on twenty mattresses and twenty feather beds when a single pea was placed under them. We are so busy examining and cherishing our own feelings that reality itself ceases to interest us very deeply.

This is the world in which Messrs Clinton and Blair - the latter of whom Peter Hitchens, the brother of Christopher, has so accurately dubbed Princess Tony - flourish. If these characters, who have long combined unctuous self-righteousness with complete ruthlessness, have been able to make fortunes from claiming to feel the pain of others, it is only because they so brilliantly distil the psychology of the age.

15

Thank You For Not Expressing Yourself

Not every devotee of reason is himself reasonable: that is a lesson that the convinced, indeed militant, atheist, Richard Dawkins, has recently learned. It would, perhaps, be an exaggeration to say that he has learned it the hard way, for what he has suffered hardly compares with, say, what foreign communists suffered when, exiling themselves to Moscow in the 1920s and 30s, they learnt the hard way that barbarism did not spring mainly, let alone only, from the profit motive; but he has nevertheless learned it by unpleasant experience.

He ran a website for people of like mind, but noticed that many of the comments that appeared on it were beside the point, either mere gossip or insult. So he announced that he was going to exercise a little control over what appeared on it - as was his right since it was, after all, his site. Censorship is not failing to publish something, it is forbidding something to be published, which is not at all the same thing, though the difference is sometimes ill-appreciated.

The torrent of vile abuse that he received after his announcement took him aback. Its vehemence was shocking; someone called him 'a suppurating rat's rectum.' He replied to this abuse with admirable restraint:

Surely there has to be something wrong with people who can resort to such over-the-top language, overreacting so spectacularly to something so trivial.

As it happens, I have myself sometimes been the recipient of such abuse: if, that is, one can be said to be the recipient of anything that remains in the virtual world alone. No subject is too recondite to provoke the insensate rage of those who disagree with the view the author has taken of it. Indeed, it sometimes seems as if fury leading to ill-mannered personal abuse and foul language is the predominant mode of disagreement in our society, at least among those who append their comments to an article that appears on the internet.

For example, I received unpleasant abuse for articles I wrote about Virginia Woolf and George Bernard Shaw. I am the first to admit that what I wrote was not emollient, indeed it strongly attacked both these figures to whom some people are strongly attached. But while I might have been mistaken in what I wrote, I do not think I am being partial in my own defence when I say that it was at least rational in the sense that it was based upon evidence culled from what they wrote. I quoted them at some length precisely to avoid the accusation of quotation out of context.

It is not necessary to repeat here what I said about them, but I shall give just one example. I pointed out that George Bernard Shaw never believed in the germ theory of disease (possibly the greatest advance in medical science ever made), regarded it as a delusion, and called Pasteur and Lister – two of the greatest benefactors of mankind, if one is prepared to admit that there can be such – impostors and frauds who had no idea of scientific method: unlike George Bernard Shaw, presumably. This was a preposterous, but not untypical, misjudgement of his, and one which he never recognised as such. Indeed, he went on re-publishing his libels on their memory until quite late in his life.

I suspect that he had that contrarian mindset that supposes that the truth must be the opposite of what everyone thinks, instead of the judicious mindset that supposes that the truth might be the opposite of what everyone thinks.

From the quality of the replies that I received, you might have supposed that I had animadverted on the moral qualities of the mothers of Latin American sons. No one ever wrote a reply (on these subjects, at any rate) claiming that I had misquoted them, quoted them out of context, misrepresented the totality of their work, overlooked their good qualities etc. I do not think I did these things, but still such replies would have been reasonable. No; I just received abuse, some of it unprintable and quite a lot of it vile.

The insults and abuse did not come from uneducated people. This is not surprising, really, because uneducated people are unlikely to care very much what George Bernard Shaw thought of the germ theory of disease; most of them have other, more practical things to think about. You have to have read Bernard Shaw to care, and these days at least, I think only university types are likely to do that.

Indeed, much of the abuse, even the vilest, came from university professors. Almost to a man (or woman), they said that what I had written was so outrageous, so ill-considered and ill-motivated, that it was not worth the trouble of refutation. On the other hand, they thought its author was worth insulting, if their practice was anything to go by. I didn't know whether I – a mere scribbler – should feel flattered that I was deemed worthy of the scatological venom of professors (not all of them from minor institutions, and some of them quite eminent).

What struck me most about these missives is the sheer amount of hatred that they contained. It was not disdain or even contempt, but hatred.

It was as if the writers had had an abscess waiting to burst, and it had burst over me. I was but the occasion, not the cause of, the discharge. But what was the cause?

These professors of hate would, I am sure, not have put their pen to paper, in the old-fashioned way, to express their feelings; they would not have written to a newspaper in the terms in which they replied to me over the internet, and certainly they would never have expected it to be published. In the days before the internet existed, or before access to it was virtually universal, I used often to receive letters through the post in response to my articles. It is not quite true that I never received abuse, but such abuse was largely from isolated cranks – the address of the newspaper in red ink or an envelope cut in two and sealed with sellotape was a clue to disagreeably expressed dissent to come.

However, for the most part criticism of what I had written was reasoned and tolerably polite. It is with chagrin that I must admit that sometimes my critics were right: I had made a mistake in fact or logic, or (worst of all) in grammar. I consoled myself with the exculpatory thought that anyone who wrote as much as I was bound sometimes to make mistakes.

With the coming of the internet, the tone of the criticism changed.

It became shriller, more personal, more hate-filled. It wasn't just that I had made a mistake, I must be an evil person, probably in the pay of some disreputable organisation or other. (There are very few of us who are not in the pay of someone, and no one is entirely reputable.)

Now of course I am not entitled to conclude from the change in the tone of criticism that I received that the internet has filled the world with hate that was not there before. It is possible that the kind of person who used to write letters to the authors of newspaper articles had fallen silent, while those seething individuals who fired off derogatory or insulting e-mails had previously been silent. But I rather doubt it.

The immediacy of the response which the internet makes possible also means that people are able to vent their spleen in a way which was not possible, or likely, before. The putting of pen to paper, to say nothing of the act of posting the resultant letter, requires more deliberation than sitting at a computer and firing off an angry e-mail or posting on a website. By their very physical nature, then, letters are likely to be less intemperate than e-mails.

The question now arises as to whether it is a good thing that people should be able now so easily to express their rage, irritation, frustration and hatred. Here, I think, we come to a disagreement between those of classical, and those of romantic, disposition.

According to the latter, self-expression is a good in itself, irrespective of what is expressed. Indeed, such people are likely to believe that any sentiment that does not find its outward expression will turn inward and poison the person who has not been able to express it. Better to strangle a new-born babe and all that.

The person of more classical disposition does not believe this. On the contrary, he believes that there are some things that are much better not expressed at all. He counterbalances his belief in the value of freedom of opinion with that in the value of freedom from opinion. He believes that rage will not decrease with its habitual expression, but rather increase with it.

By now it should be clear which of these two viewpoints seems to me to be the more accurate. The habit of not containing your rage is likely to lead you to easily provoked enragement. And, as almost everyone knows who has taken the trouble of self-examination, there is a great deal of pleasure to be had from rage, especially when it supposes itself to be in a righteous cause.

'So,' I hear an imaginary interlocutor say, 'you are in favour of censoring the internet.'

I know that this is how some people will respond, because when I argue that the balance of the evidence suggests that children who grow up with a mental diet of violence on electronic media are more likely themselves to become violent than those who do not, someone will almost always pipe up 'So you are in favour of censorship.' In vain do I point out that there is no strictly logical connection between the causation of violence by electronic media and censorship, because one might think that censorship was a worse evil than any resultant violence; after all, no one wants to ban cars because there is no speed at which they are entirely safe and without the possibility of causing a fatal accident. There is no need to blind ourselves to the undesirable effects of violence on television, films, etc. just because we don't like censorship. Indeed, to do so is a form of voluntary self-censorship, perhaps the most insidious kind.

So it seems to me at least possible that easy access to public self-expression tends to make people more bad-tempered and ill-mannered than they would otherwise have been. It releases people from inhibitions, and allows them to breach psychological barriers. Even wit suffers, for it is far easier to insult than to think of a really damaging, but amusing, witticism. To write to Professor Dawkins that one feels 'a sudden urge to ram a fistful of nails down your throat' is easier than to explain succinctly why he is wrong, if he is wrong.

Moreover, the fact that one can vituperate using a virtual rather than a real address promotes such verbal intemperance.

I don't suppose there is an easy solution to this problem; that is, if it is a problem. The auguries are not particularly good if it is also true, as it is in my experience, that professors of literature are among the worst offenders. If those who teach youth are unable to control themselves, and to keep their disagreement within the bounds of common civility, what can we expect of youth itself?

Perhaps, on thinking of the limits to what we can rightfully say, we should ever recall the lines of the immortal Swift, towards the end of his Verses on the Death of Dr Swift:

Yet, Malice never was his Aim;
He lash'd the vice, but spar'd the Name.
No Individual could resent,
Where Thousands equally were meant...
For he abhorr'd that senseless Tribe,
Who call it humour when they jibe...

The problem is, of course, that of no writer of the century in which he wrote was this less true than of himself.

16
Steel Yourself

T here is nothing more melancholy than a steel town without the steel. Even at their most prosperous, such towns are not often lovely; but at least they can then generate a sense of individual purpose and municipal pride. When the steel departs, these two things depart with it. As it happens, I have had occasion recently to spend a few weeks in two steel towns in Great Britain from which the steel has fled to more welcoming climes abroad. One was in the North of England, and the other was in Wales; both were grim almost beyond description, though there were slight differences between them.

The one in Wales, for example, still has architectural traces of its strongly Baptist past. There were many chapels, of surprising if forbidding and soot-covered grandiosity, still standing, though now not much attended. I confess that my attitude to the Welsh Baptist form of Christianity is somewhat ambivalent.

On the one hand, I can see that it created and sustained a sense of community where life was generally very hard and without many social amenities. It gave to people a strong sense of morality in a situation in which only such a sense could have made life bearable. Licence is no friend to the poor.

On the other hand, the aesthetic aspects of the religion do not please me, its strong choral tradition notwithstanding; it was dismally puritanical, and the morality it inculcated and enforced by social pressure was often a narrow and intolerant one, without any sense of irony. I think this weakness (a weakness integral to its strength) is best summarised by the only joke I know about Welsh Baptist religion:

> A preacher finishes his sermon. A young man in the congregation puts up his hand to ask a question.
> 'Preacher,' he says, 'is it all right to have sex on Sundays?'
> The preacher thinks for a moment.
> 'Yes,' he replies. 'So long as you don't enjoy it.'

Of course, to appreciate the joke fully it has to be told in what is to me the beautiful and poetical Welsh accent which, however, is very well suited to a kind of unctuous religious hypocrisy.

Like everywhere else in Britain, though, religion is now dead or dying in Wales (except that the town of which I speak is now vying to become the location for a large 'Halal industries' site, though at the moment it has very few Moslems, which will supposedly create 1500 badly-needed jobs).

No doubt people will go on arguing forever about whether the kind of religion that was once predominant in the town was a cause or a consequence of its former industrial greatness, that is still to be glimpsed in the grandiloquent municipal buildings that stand out in the town like a few rotten teeth in an otherwise edentulate mouth; but it is hard to imagine that the fortitude, self-control, self-discipline and self-respect that it inculcated in the population was of epiphenomenal importance only.

By contrast, the town in the North of England had once been dominated by a magnificent Fifteenth Century church, originally Catholic, of course, but now Anglican. It was totally irrelevant to the life of the town, if life is quite the right word for what went on there. The vicar was an amiable and kindly man – it was written on his face – but these were not the times, nor was this a place, propitious to the exercise of amiability and kindliness. Many of the buildings around the church, some of them hundreds of years old, were boarded up against the inevitable vandalism; there was no other possible purpose for them now.

Even more distressing to me was the Church of England's complete disregard for its own aesthetic heritage. For many years now a host of snivelling cowardly bishops have sought to do the impossible, curry favour with the liberal intelligentsia, by abandoning the magnificent Book of Common Prayer for a version that reads as if it were written by the Ministry of Transport of a country in the throes of reconstruction after a devastating war; but what applies to the liturgy applies also to the physical fabric of the churches.

With complete disregard for the aesthetics of this church – a magnificent monument in a wasteland of man-made hideousness - cheap modern furniture had been installed, and even (in place of a lady chapel) a kitchenette, complete with plywood cladding, used for the doling out of tea to lonely old ladies.

Please do not misunderstand me: I am not against the doling out tea to lonely old ladies; indeed, I am much in favour of it. But I do not think that a Fifteenth Century church is the right place for it, especially if the interior of the church has to be spoiled in the process, and the fact that the Church of England thinks that this is all right accounts in some part for its demise. The kitchenette was visible and obvious evidence of the Church's lack of belief in transcendence, in anything other than the most earthbound of values. (The ancient tombstones had also been removed from the grass around the church for reasons of public safety.)

The ex-steel town in Northern England had been selected by the government, or by some department thereof, as a place of settlement for asylum-seeking refugees, mainly Kurds from Iran and Iraq. These young men – they were overwhelmingly young men – had made their way across dangerous and hostile territory, often by very chancy means, to reach Britain. I know from experience of talking to them that it is not easy to arrive at an estimate of their true motives for coming: but whatever those motives might be, their initiative and willingness to take risks can scarcely be doubted. If allowed to be, they could be an asset to the country that received them.

Instead of which the official policy was to turn them as quickly as possible into welfare dependants. Unable either to prevent them from coming or to deport them once they have arrived, British officialdom in its wisdom has decided to prohibit them from working, and to enforce this prohibition it has selected places like this ex-steel town, where unemployment is near-universal, as their enforced place of residence. So there they congregate, quickly turned from adventurous and eager young men into dispirited idlers, mere *habitués* of billiard halls and consumers of pornography on the public library's computers (or as near to pornography as library's system would permit): that is, when the most enterprising among them they did not become traffickers in something or other.

In both towns, the only economic activity is the administration of poverty and the recycling of government subventions, usually through supermarkets and charity (thrift) shops. Even the charity shops are in effect governmental because most of the larger charities in Britain have been nationalised, their most important donors now being the government. I enter these shops – the Germans have a saying that never fails to come to mind as I do so, namely 'It smells of poor people here' – because they always have a few books for resale, overwhelmingly the trashiest of trashy novels, but usually (and unaccountably) with an academic tome among them at a knockdown price, for example Herbert S Klein's *African Slavery in Latin America and the Caribbean*.

If anyone doubts the existence or reality of a dependency culture, he should visit one of these two towns. They are East European communist towns with a bit more consumer choice, but not much the better for that, and in some respects worse, in so far as there is less intellectual ferment in them.

The people do not walk so much as trudge, plastic bags hanging from them like heavy fruit. They are grey-faced, bowed-down, prematurely aged, arthritic before their time. An astonishing proportion of them need (or at any rate use) walking sticks from their thirties onwards. Many of them are enormously fat, and one can imagine them completely immobile by the age of sixty. The small children – overwhelmingly illegitimate, of course, for more than half of children born in Britain are now illegitimate, and the poorer the area (except for Indian and Pakistani immigrants) the higher the proportion – are devoid of the sweetness of young childhood, instead having a fixed look of malice on their faces by the age of three. Ferret-faced young men, attired in international ghetto costume, often with a hood, stand around talking to one another, at least a third of their words being 'fuck' or one of its cognates. The young women are all highly sexualised without being in the least alluring. Their fate is to have children by more than one of the ferret-faced young men.

Hopelessness, indifference, apathy is everywhere, omnipresent like the gases of the atmosphere. No Indian or African slum has ever affected men in the same way: this is far, far worse. Energy is dissipated before it is expended, as if by some kind of magnetism. The people are not starving – if anything, the problem is the reverse – nor are they living in physically intolerable housing conditions, though their houses are depressingly ugly. That so many are festooned with satellite dishes is a bad sign: where satellite dishes are many and prominent, the people are bored and listless. Litter lies everywhere and many people do not clear it even from their own front yards, preferring to wade their way through it to their front doors.

It astonishes me, however, that when I speak to the people here – posh voice, obviously an emissary from another world, if not from another universe altogether – I am responded to not with hostility, but with smiles (though I inwardly remark on the terrible dentition), kindness, cheerfulness, and helpfulness if, for example, I want directions. I am not sure I could live their life and talk to a stranger so politely.

This underlying decency makes me all the sadder. I think of the words of Edward, Prince of Wales, when he visited South Wales at the height of the Depression: 'Something,' he said, 'must be done.'

Yes, but what? Certainly, pity, with its almost inevitable leaven of condescension, is no answer.

The situation is this: the people, for the most part, are not well-educated and they have no skills. The next generation will not be well-educated either, because the state educational system steadfastly refuses to teach. But unskilled labour in factories is unavailable and will never be available again, at least not without protectionism that would wreck the world economy as a whole.

Thanks to the subventions they receive, and no doubt to the loss of the work ethic or habit, their labour is unlikely ever to be worth as much to an employer as he would have to pay them to make it worthwhile for them to go to work. After all, as things stand, their rent is paid, their local taxes are paid, their schooling and health care are paid, they have to make no contribution to ensure they will get a pension, and their bus fares are paid. Their one responsibility is to stretch out the cash-subvention that they receive in such a way that they can eat, smoke, drink and watch television.

They cannot start little businesses of their own, of course, because of regulations: regulations ostensibly for their own benefit, for the sake of their health and safety. In any case, if commerce were free, if the entry price to it were much lower than it is, the most important economic activity of these towns – the administration of their own poverty - would be severely threatened. There is now a nomenklatura class, as a glance in the staff car-park of any welfare institution will prove. So everything must remain the same, forever, at least until a general collapse.

In one of the towns the local newspaper enjoined its readers to vote for the baby of the year (illegitimate, of course, but bonny for the time being). Here, at least, was something worthwhile to vote about, unlike the forthcoming general election in the country with its choice between Labour and Conservatives, between Mr Brown and Mr Cameron, between Tweedledum and Tweedledee.

17

The New Faith, Hope and Charity

C onfucius said in the *Analects* that the first thing he would do on coming into power was to ensure that things were called by their proper names: for if they are not, what confusion follows!

But confusion does not arise from poor nomenclature only. Correct naming is a necessary but not sufficient condition of clear thought. Unexamined premises and false assumptions are a fruitful source of bad thought and worse conclusions.

There is no better time to study bad argumentation than during an election in a western democracy. No falsehood is too false to be suggested, no truth too evident to be suppressed, no *non sequitur* too illogical to be enunciated, by a candidate during one of these contests. They cut words adrift from their meaning, leaving only a connotation behind, like the grin of the Cheshire cat. A fastidious man, with no experience of living under any other political system, is likely to react with disgust.

The current election in Britain is a fine example of the genre. It is particularly distressing to me because the shortcomings of no country affect one as deeply as those of one's own. And the candidates in this election seem perfectly to exemplify the weakness of contemporary British culture: frivolity without gaiety and earnestness without seriousness.

Against my principles and practice of forty years, I allowed myself
to be persuaded by friends to watch a so-called debate between the three
principal candidates in the election. Of course, a three-way debate is an
inherently unsatisfactory thing, like a dog with five legs, or a war on two
fronts; but I had no confidence that a debate between any two of them
would have been better or more illuminating.

In the event, the 'debate' was more like a trialogue of the deaf;
when one of them made an outrageous statement, as he often did, the
others said nothing by way of refutation. For example, asked about the
advantages (or otherwise) to Britain of its membership of the European
Union, one said that the European Working Time Directive – a regulation limiting the amount of time a paid employee can be asked to work
in a week – had given us paid annual vacations, while another indulged
in an almost lyrical description of how European co-operation had permitted the dismantling of an international ring of paedophiles.

The third of the candidates said nothing to contradict these outrageous and totally irrelevant statements. One would have supposed that,
until the European Union came into being, no one had ever hit upon
the idea of annual paid holidays; or that no international police co-operation had ever happened before the establishment of the European
Union. Moreover, if the best thing that could be said about the giant
(and vastly expensive) European bureaucratic apparatus is that it once,
in over thirty years of British membership, had facilitated the dismantlement of a paedophile ring, at the cost of untold billions of dollars, it was
not very much. I am, of course, as much against paedophilia as the next
man; but there is a limit to the number of tax Euros that even I am willing to devote to the dismantlement of one paedophile ring every third
of a century.

This was all demagoguery of the purest strain, of course. When
one of the candidates (the current incumbent of the top post) said that
an enormous proportion of our trade was with the European Union, no
one asked him asked him what proportion of it would cease if Britain
were not part of the Union. After all, trade is conducted on the basis of
mutual advantage, mainly – though perhaps not entirely - economic;
and in a world in which tariff barriers are of less and less significance,
what special or specific advantage is to be derived from the membership
of an association whose main activities are bureaucratic, regulatory and
prohibitionist? Perhaps there is such an advantage, but no one asked the
incumbent to specify exactly what it is.

As for the paedophile ring assertion, it was beneath contempt. UNICEF recently found that Britain was among the very worst countries in the western world in which to be a child; and while I would not normally give much credence to the findings of this particular organization, in this instance it was so obviously and self-evidently correct that I am happy to cite its report as evidence. The danger to British children comes not from paedophile rings, but from British parents; indeed, many British parents so hate or neglect their children (except when they are babies) that it is a wonder that they bother to have them in the first place. More than half of the children are illegitimate, and illegitimate children are much more likely to be the victims of paedophilia than legitimate ones, a fact well-known by now to all British parents, who nevertheless continue to indulge themselves; many more British children have a television in their bedrooms (to shut them up) than a father living at home; a few years ago, 36 per cent of them never ate at a table with another member of their family, a proportion that has almost certainly risen. In the circumstances, then, to cite the dismantlement of a paedophile ring was to appeal in the most blatantly demagogic way to the anti-paedophile hysteria in Britain, an hysteria that is itself the product of a justifiably guilty conscience about the way many children are brought up.

Of course, I can quite see that accusing one's electorate of gross irresponsibility (or worse) in its child-rearing practices might not be the best way to win an election; but in that case, at the very least, the subject of paedophilia should have been avoided by the candidate, especially as it was he who brought it up.

Towards the end of the 'debate,' if that is what it was, a woman in the audience aged 84, well-preserved and well turned-out, asked the three leaders whether they thought that the $80 state pension that she claimed to live on was enough, especially as she had brought up four children and worked hard all her life.

All of the candidates agreed that it was a terrible thing that she should live on $80 a week; none had the courage to face her down and tell her that what she had said was obviously a misrepresentation of reality, if not an outright lie. After all, if you gave an 84 year-old woman $80 to live on, without any other source of income or subvention, she would not (at least in British circumstances) live very long, let alone look well-preserved and well turned-out. Clearly she had some other source of income, whether it was from private sources or from public subventions. To give one small illustration: when I am in England, I live in a small house whose local tax bill is $50 a week; surely she did not pay $50 from her $80 a week in local taxes, leaving her but $30 to live on?

Now if our three leading politicians had not the moral courage to confront one old lady who was misrepresenting the reality of her situation, deliberately or accidentally as the case might be, what likelihood was there that they would be honest about the very serious problems that confront the whole country?

Again, one could quite see that the incumbent might not be very keen to be honest about these problems, since he was so largely responsible for bringing them about in the first place. He permitted cheap credit, thus encouraging the asset inflation that reassured the middle class that it was effortlessly growing rich, while simultaneously and enormously expanding public expenditure and the public payroll. Duped by the appearance of ever-increasing asset values, millions borrowed against these rising values in order to fund current sumptuary expenditure, imagining themselves, therefore, to be living well. When the music stopped, the debt, both public and private, remained, but the value of everything else had melted away. In essence, things were not very different on the other side of the Atlantic, though the public money was used, or wasted, somewhat differently.

You might have thought, then, that nothing would have been easier than to criticise the incumbent on his economic record. On the contrary, there was no such criticism during the debate, I think for two reasons: first, reflection on the facts might suggest a necessity for a retrenchment, a downward adjustment of the general standard of living, which is not what any electorate likes to hear, even – or especially – if it is the truth. Second, it suggests that, while the government and the banks have been irresponsible, so has the general population. Much of the latter has behaved like children who cannot resist chocolates, gorging on them until they feel (and are) sick. Perhaps it is not surprising if politicians seeking votes are reticent about the folly of the voters; and it would certainly take courage to confront the population with the truth. Courage is precisely what modern politicians lack; and they would far rather be in power than be right. Truth is the first casualty of an election.

It is impossible to avoid election propaganda, however hard one tries to do so. Recently a propaganda sheet for the Liberal Democrats, who have emerged as the third force in British politics, was pushed through my letter box. This was unintentionally revealing, for it demonstrated (unconsciously) how far people now exist for the state rather than the other way round.

It showed the local candidate with the party's candidate finance minister holding a huge simulacrum cheque, written out to 'The Tax-payer' for the sum of $15,000 – the amount of income that the Liberal Democrats suggest each person should be allowed before his income is taxed. This is a considerable increase on the amount now allowed.

All to the good, no doubt: but there is something rather disturbing, and indeed sinister, about this way of putting things. It suggests that it is the government that allows or grants the people money, not the people who allow or grant the government money. To refrain from taxing is not giving money away, it is to avoid appropriating money from its original owners. If a mugger in the street were to return us a couple of dollars from what he has taken from us for our bus-fare home, we would not consider that he has been generous or 'given' us anything, even if he makes a whole ceremony of the return of the two dollars.

The fact that the matter can be presented in the way that the Liberal Democrats presented it in their election propaganda, almost certainly without any public criticism let alone protest, reveals how far we have become, and expect to be, creatures of the government. We have come to accept that the first call on our money is taxation, and that any that is left to us is by grace and favour; being allowed to keep what is ours has become, in effect, a donation to us by our political masters. On the rare occasions when a tax is reduced in Britain, the Chancellor of the Exchequer (the Finance Minister) is said to have 'given money away.'

In return for this appropriation of our funds by politicians, they offer us all kinds of benefits, and it would be dishonest not to acknowledge that we do indeed receive some or many of them, and that, once we have received them, we are reluctant to relinquish them, however unsustainable in the long term they might be. I am not one of those that believes that Man naturally desires freedom, at least if by a desire for freedom is meant a desire that automatically trumps all other desires and is prepared to take the consequences. What our politicians have learnt to hold out as the prospect before us, like a mirage in the desert, is the greatest, most sought-after and least possible freedom of all, the freedom from consequences.

The idea of infinite benevolence has been transferred from the deity to the government, nowhere more successfully than in the minds of the governors themselves. Illusion, Benevolence and Power are our new Faith, Hope and Charity.

18
The Machine

Now that the world's economic centre of gravity has moved decisively towards the east, and eastward the course of empire takes its way, some western economists have discovered that economic growth is not all that it was once cracked up to be. They have discovered what the moralists of all ages have known, that happiness is not proportional to levels of material consumption, at any rate once absolute poverty has been surpassed.

For example, Robert Skidelsky, the biographer of John Maynard Keynes, says so. As literary theorists might put it, 'the subtext' of this is that the declines in per capita GDP experienced in European countries as a result of the economic crisis are nothing much to worry about, since those countries have long since reached the level of consumption at which increases add nothing to happiness. A decrease of ten per cent, say, takes us back to the levels of about four, five or six years ago, when nobody was remarking on the terrible suffering caused by the poverty of the mass of the population.

On all this I am, as on many subjects, of two minds. Moreover, my practice does not always perfectly reflect my thoughts of the moment.

On the one hand, I am fully aware that my own personal happiness has had very little to do with my economic circumstances (I speak as one who has never had to worry much about where his next meal was coming from). If I look back on my life, some of the happiest times

I have known have been when I had least, not most, consumer choice. To eat what there is rather than what you choose is to be relieved of a distracting dilemma. In a street of restaurants, and hungry, I have often been like Buridan's ass, unable to make a decision between them, and fearing to make a mistake: a mistake which, if made, would not be so very terrible.

One has only to observe a streetful of shoppers on a Saturday afternoon to understand the futility of consumption as a path to happiness. What, exactly, are they looking for? It is rarely that one sees among them a look of ecstatic happiness that tells you that they have at last found what they wanted all their life to find. I think if I found a Vermeer in a junk shop (my fantasy from an early age) I should be genuinely happy; more recently, I would make do with a mere Thomas Jones.

As it happens, I did once find a Thomas Jones, but unhappily it was not in a junk shop. It was in a fine art gallery and cost $400,000, which I did not have. In case you are wondering who this Thomas Jones is, or rather was, he was a Welsh painter of rather ordinary landscapes, active in the second half of the Eighteenth Century, who, however, suddenly painted a few pictures of transcendent beauty and genius: exquisite depictions of Neapolitan roofs and walls, pictures of such serenity that they gladden not just the eye but the soul. It is the very ordinariness of what is shown to be of transcendent beauty that is so inspiriting; and it starts you on the search for all the beauty, previously unnoticed because taken for granted, around you. Never again will you look on a wall with the same dull inattention.

Suffice it to say, however, that most Saturday shopping is not for a Vermeer or even a Thomas Jones; it is for objects the pleasure of whose possession will mostly last no longer than the bloom on a grape and, being passed, will be immediately succeeded by another futile search for another joyless possession. The good consumer never learns.

On the other hand (what was on the one hand having being stated four paragraphs ago), I must confess to not being an economic ascetic. I rarely turn down offers of work for money, and am pleased if I am paid more rather than less. My taste in food, when I have the choice, runs to the expensive rather than to the cheap. I still covet antiquarian books, though biology ensures that I can no longer enjoy them for many years (and, despite the general decline in our civilisation, level of education, advance of the internet etc., antiquarian books are still very expensive, suggesting that there is a market for them). Though by no means rich, I am several orders of magnitude away from mere subsistence. Where my standard of living is concerned, I cannot claim to be a child of nature, a Gandhian who weaves his own cloth and grows his own lentils. (Of course, it was said to have cost a fortune to keep Gandhi poor.)

Like most people, also, I am depressed by bad economic news. For example, when the number of new cars bought last month is reported to have declined I feel gloomy, even though I think there are far too many cars already, that they have had a baleful aesthetic effect on much of the world, and that – in large cities – they are as often a source of frustration as a means of quick transportation. (It is said that traffic in central London moves no faster than in Victorian times.)

When I see the price of gold rise, I regret that I did not follow my instinct and buy it when it was much cheaper (I get no equal and opposite pleasure when its price falls.) A rise in the price of gold means a loss of confidence in the preservation in value of everything else: the everything else of which I own some. Poverty and hunger are staring me in the face! What a reward for having lived within – no, below – my means for many years!

Moreover, I am a consumer just like everyone else, no worse, perhaps, but certainly not much better.

Last week, for example, we bought a large new refrigerator. Why? Our old one still worked perfectly well. It was small, but we didn't need anything larger; it was perfectly adequate to our needs, from the preservation of food point of view. We bought a large new fridge because we had to bend down to get anything out of the old one, and we decided that we didn't want to do that any more. We wanted an eye-level refrigerator; and in order to have one, we had to buy something much larger than we needed. We rationalised our purchase by telling ourselves that we were growing older, that soon we might not be able to bend, it was better to be prepared in advance for the difficulties of old age than face them as an emergency, etc., but really our purchase, quite unnecessary, was merely whimsical, at best to overcome a very minor inconvenience.

What did we do with our fridge? These days you cannot give away what would have made Louis XIV green with envy. You could, for example, go down the road shouting 'Free fridge! Free fridge!' and find no takers. Indeed, you might end up in an asylum. So we took it to the municipal wasteground of such things, where people dispose of what they no longer want.

Although our town is small, there was enough there to equip scores of households. The attendant told us that they did try to give these things away, after having tested them for safety, but it was not easy:

there were more discarded goods than people to need them. I am no environmentalist, but still I could not help but feel that there was something amiss in all this.

I like grapefruit juice in the morning (I detest orange) and am too lazy to squeeze it for myself. I think, no doubt from vanity, I have better things to do with my time. If I buy three large cartons of freshly-squeezed grapefruit juice it is 20 per cent, per cubic centimetre, cheaper than buying one. Like the vast majority of human beings, I can't resist a lower price, even when the lower price will make a saving that itself will make no difference to my life. So I buy three cartons of grapefruit juice instead of one.

The problem is that the grapefruit juice won't keep long enough if I drink it at the rate that I normally would; so, in order to save between $1 and $1.40 over a period of, say, two weeks, I drink every day more grapefruit juice than I otherwise would or, truth to tell, than I really want.

No wonder that people these days are so fat.

Whenever I think about these things, my mind returns to the title of a short story by E M Forster, unusually for him in the genre of science fiction: *The Machine Stops*. Set in the long-distant future, the story tells of people living underground who are utterly dependent on a vast machine for ventilation, distribution of food and sustenance etc., a machine that no single person controls or even understands. The story imagines what it will be like for these people when the machine grinds to a halt and no one can repair it. Divorced for ages from any form of primary production, they will die, and die horribly.

Our consumer society is like the machine in the story. My wife and I (principally my wife) have grown a little produce, it is true, but it took ages, cost a fortune and wouldn't have kept a hamster alive for a week. Autarky is not really for us, or even primitive exchange and barter with our neighbours. And so, to ensure that the machine does not stop, we have to do our duty as consumers. In other words, we (humanity) are the creatures of what we ourselves have created.

This is not as much of a criticism of the consumer society or economy as might be thought. All judgment, said Dr Johnson, is comparative, and the correct standard of comparison in this instance is not the perfect but impossible, but the imperfect but possible. Our choice is always between evils, more or less partial as they may be.

I do not believe that it is possible to devise a system in which we all live well, to a standard that will satisfy us, and in which there is stable equilibrium: that is to say a state in which we all have exactly what we want, but no more than we want, and in which no one tries to persuade us to want more than we currently want. The machine needs expansion, not stability.

The real alternative is not perfect equilibrium, therefore, but real, and in the end absolute, poverty. This practically none of us is willing to countenance; none of us wishes to return to a back-breaking subsistence. Our bargain is a Promethean one.

That is why western economists who now say that happiness is not proportional to consumption once a certain level has been reached are saying something that is true but beside the point, except for those few individuals who are willing to limit their desires. Though they are admirable as people, those self-contained people, and indeed generally more interesting than those who see the meaning of life in the consumption of quickly-discarded goods, we must hope that, in the long-term, there are not too many of them: for if there were, the machine would stop.

On the other hand, having lived beyond our means for many years, in essence on money borrowed from the people who have produced all the things that we don't need and that haven't made us any happier, it is to be hoped that there are many among us those who will be able to reduce their sumptuary expenditures temporarily, until things come back into proper economic balance. Once that balance is achieved, of course, they will need to expand their desires again. From what I know of my fellow-beings, I think there will be no real difficulty in this. As for me, I shall continue to do my duty, and drink more grapefruit juice than I want.

19
Of Snobbery and Soccer

A n acquaintance of mine, whose opinions I generally respect, once said that snobbery is a vice, but a very minor one. I am not so sure.

Like many phenomena, snobbery is easier to recognise than to define. The definition of a snob in *The Shorter Oxford English Dictionary* is inadequate. *The Penguin English Dictionary* does much better. It defines a snob as 'Someone who tends to patronize or avoid those regarded as social inferiors; someone who blatantly attempts to cultivate or imitate those admired as social superiors; someone who has an air of smug superiority in matters of knowledge or taste.' The same dictionary defines 'inverted snob' as one 'who sneers indiscriminately at people and things associated with wealth and high society.' One possible derivation of the word snob is from the Latin *sine nobilitate*, without nobility.

I doubt whether there is anyone in a modern society who is entirely free of snobbery of some sort, straight or inverted. After all, everyone needs someone to look down on, and the psychological need is the more urgent the more meritocratic a society becomes. This is because, in a meritocracy, a person's failure is his own, whether of ability, character or effort. In a society in which roles are ascribed at birth and are more or less unchangeable, failure to rise by one's own achievement is nothing to be ashamed of. To remain at, or worse still to sink down to, the bottom of the pile is humiliating only where a man can go from log cabin to White House. Of course, no society is a pure meritocracy and none allows of absolutely no means of social ascent either; thus my typology is a very rough one, and is not meant to suggest that there is ever a society in which the socially subordinate are perfectly happy with their lot or are universally discontented with it. But it does help to explain why justice, of the kind according which everyone receives his deserts, might not necessarily conduce to perfect contentment. It is obviously more gratifying to ascribe one's failure to injustice than to oneself, and so there is an inherent tendency in a meritocracy for men to perceive injustice where none has been done.

It is not altogether surprising, then, that small slights are often felt far more grievously, and burn for longer in the mind, than large or gross injustices. A burglary is more easily forgotten than a disdainful remark or gesture, especially one made in public; one might consider this foolish, but it is irreducibly so.

That is why snobbery, when openly expressed, is so hurtful and dangerous. Even quite mild people become furiously angry, sometimes to the point of violence, when too clearly disdained. To let people know that you look down on them, *ex officio* as it were, is the surest way to provoke their antagonism. By contrast, exploitation (within quite wide but not infinite boundaries) is relatively easy to tolerate.

The antagonism that European colonialism evoked in Africa, for example, was caused more by the evident disdain of many colonialists for the local population than by grosser exploitation. Of course, in some instances the exploitation was so gross as to provoke rebellion; but by the end of colonial rule, when antagonism to it was at its most popular and widespread, the grosser forms of exploitation had been eliminated. Moreover, antagonism to colonial rule was as great in countries which clearly benefited from it economically as in those which it did not. Even economic retrogression in the post-colonial era did not result in calls for a return to the palmy days of colonialism: for no one likes to be an inferior in his own country merely by virtue of having been born in it. Colonialism was experienced as snobbery incarnate, institutionalised disdain, and therefore disliked intensely by those who experienced it.

Knowing the dangers of snobbery, however, is not quite the same as eliminating it from one's own heart and mind. I admit that, in the inner recesses of my being, I am a fearful snob. For example, I feel nothing but contempt for people for whom sport is important. This is particularly pertinent at the moment, because the greatest sporting event in the world by far, the football (soccer) World Cup, is taking place in South

Africa as I write this. There could be no greater snobbery than to feel contempt for the hundreds of millions of people world-wide for whom this event is of consuming interest. When bread is assured, circuses fill men's minds.

Nowhere has this been more so than in France, where a veritable crisis has been caused by the utter failure of the national team. That team played lamentably badly, failed to win a single match, lost against the most mediocre opponents, was eliminated from the competition at the first hurdle, and worse still behaved abominably.

In 1998, the French team won the World Cup and there was a burst of national euphoria as a result. The team of 1998 was composed of *blancs, beurs, noirs* – that is to say, whites, Arabs, blacks – and this was taken, briefly, as evidence of the success of France as a multicultural and multiethnic society. Huge crowds greeted the successful team as it paraded in the modern equivalent of a Roman triumph. Preposterous triviality could go no further.

Twelve years later, when the French team lost miserably in the same competition, the opposite sentiments were widely expressed, at least in the newspapers and on the air. The team was now predominantly black and Arab; anyone who knew France only through its national football team would place the country somewhere between North and Equatorial Africa. One prominent white in the team, a spectacularly ugly and thuggish-looking man, so ill-educated that he could barely string a few words together, let alone a sentence, in his native language, had converted to Islam. Another white in the squad, a blonde Breton who was notably better-educated than his colleagues, had to be excluded from the team because none of the others would co-operate with or pass the ball to him.

When the Marseillaise was played before a match started in which the French were to play, the team refused to sing it or accord it any respect. While it is perfectly normal for many Frenchmen not to know all the words – which is probably as well, since they are horribly bloodthirsty, and include the hope that the impure blood of aristocrats may irrigate the ploughed furrows of the peasantry – almost all know at least the first four lines. The players appeared to be expressing their disdain for the country they supposedly represented and that had enabled them to become multi-millionaires by the age of 20. At the root of their resentment would not be injustice, but remembered slights, real or imagined.

A player named Anelka then insulted the team manager because the manager criticised the performance at half-time. The words used

by Anelka were so gross that I will not translate them; the manager excluded him from the team and sent him home. Outraged by this assault
on their inalienable right to freedom of speech, the team went on strike
and refused to train for a day. It was hardly surprising that even sixth-
rate teams were able to beat them.

Whereas the victory in 1998 was taken as proof of the success of
French society, the defeat in 2010 was taken as proof of precisely the opposite. How was it that the country had raised up a generation of resentful, ill-mannered, ungrateful and thoroughly spoilt youths, who weren't
even very good at what they had been paid enormous sums to do? (One
of the better-behaved and more dignified of them, a man called Thierry
Henry, is paid more than $20,000,000 a year, before his advertising and
publicity revenue.)

Of course, if the team had been successful, if it had repeated the
success of 1998, no one would have raised these questions, and euphoria
rather than depression would have been the mood of the moment. As it
was, the team was the best propaganda possible for Jean-Marie Le Pen,
of the *Front National.*

A parliamentary enquiry is to be held about the state of French
football; the president himself has expressed his concern. Many people
have said that debacle reflects the state of French society. Against all this,
one writer in *Le Monde* did manage to point out in a short article to the
effect that football is only a game, after all, and the whole spectacle a
trivial one; but the opposite view of its importance prevailed.

I confess that I was surprised by how the French showed themselves as stupid about football as the English. It is true that the behaviour
of the French team was used as a metonym for the horrible, resentful
culture of the suburban housing projects that surround every French
town of any size; but it was hardly necessary for the French team to have
behaved so badly or to have lost for the latter to be widely known. It is
also true that if you compare the faces of the English football team of,
say, the 1950s with those of the team today, you will see the decline in
civility of English society as a whole. But what really mattered to people
in France was victory or defeat in the sporting contest, not the state of
society. Football was more important to them than anything else, and a
victory – or at any rate, a more dignified defeat – would have anaesthe-
tised their thoughts about the country's social problems.

It seems to me very odd, and not at all reassuring, that a country
such as France, with a practically unrivalled history of achievement in
all the major fields of human endeavour, should have been precipitated

into an orgy of self-examination by something as trivial as a failure in a football competition, when it is utterly indifferent to questions of incomparably greater importance: for example, why it is completely incapable, after a continuous and millennial history of wonderful architecture, of erecting a decent building, one that is not an eyesore? (It is not alone in this, of course.) I have never seen this question so much as raised, let alone answered, though I do not think any reasonably alert person could drive through France without asking it.

The decerebrating effect of football (and no doubt other sports as well) is illustrated by a story that my French brother-in-law told me recently. A couple of months after France won the World Cup in 1998, he went to Tibet. He went to a Buddhist monastery that was two days' hard trek from the nearest road. There he met young novices, some of whom spoke a few words of English. They asked him where he was from and he told them.

'France,' they said. 'World Cup. Zidane.'

The glory and civilisation of France was thus reduced to eleven men on a field successfully, and admittedly with great skill, kicking a ball about. Zidane, incidentally, was a player of Maghrebian descent, the great hero of the 1998 competition and a man who looks considerably more intelligent than any of the players today; he blotted his copybook slightly when he head-butted another player, an act that he explained by saying that you can take a boy out of a slum, but you can't take a slum out of a boy.

On the subject of football, I am a snob. I do not detest the game as such, for I accept that it can be played with skill and achieve a kind of beauty, but rather the excessive importance attached to it by millions and hundreds of millions of my fellow beings. Try as I might to expunge the thought from my mind that this enthusiasm is a manifestation of human stupidity, I cannot.

Of course, we are all of us snobbish about something or other; the important thing is to control ourselves and not express our snobbery openly, so that we do not give offence by it. I am therefore always careful to disguise my contempt for enthusiasm for football from enthusiasts. Besides, if I were to reveal it, they might hit me.

20

Destructive Preservation

T he one thing that many environmentalists seem not to care about is the environment. By this I mean its visual appearance. They would happily empty any landscape or any city of beauty so that the planet might survive. Like the village in Vietnam, it has become necessary to destroy the world in order to save it. And, of course, destruction of beauty has the additional advantage of being socially just: for if everyone cannot live in beautiful surroundings, why should anyone do so? Since it is far easier to create ugliness than to create beauty, equality is to be reached by the former rather than by the latter.

The indifference of environmentalists to aesthetic considerations was illustrated by a friend, who kindly forwarded to me a brochure about a fully ecological house, erected (or assembled, since it was prefabricated) in the centre of Paris in front of Haussmann-style buildings. Needless to say, it completely destroyed the harmony of the surrounding townscape.

It looked like a three-dimensional Mondrian, all boxes and bright colours. Inside, it was more a laboratory than a home, the kind of sterile environment necessary for in vitro fertilisation. However much it might have been heated by the sun, it lacked warmth. It was a proper place for androids, not for humans.

The brochure claimed many advantages for it, not the least of which was that the residents could monitor their energy consumption electronically hour by hour, minute by minute, in order to minimise it. Thus they could ensure that they never forgot their own impact on the environment, and were never totally free of anxiety about it. What the saving of their souls was for the ancients, saving of electricity has become for the moderns.

No consideration was given in the brochure to such questions as the harmonisation of new houses with the pre-existing townscape or landscape, or how these cheap and gaudy constructions would look after a few years of wear and tear; but the smallness of the houses was vaunted as an enormous social advantage. There simply was not enough room, not enough land area, said the brochure, for everyone to occupy as much space as he wanted.

This was an odd claim, because the house was by no means as efficient in concentrating the population as – the very Haussmann-style buildings in the front of which it was assembled, which manage so marvellously to combine elegance, grandeur, human scale and density of population, and which are now so desired and desirable as places to live that they have become too expensive to buy for anyone who does not already own part of one. Oddly enough, no one has ever suggested building as Haussmann did, albeit with such energy-saving devices as ingenuity might supply. The past is the one thing we don't want to learn from, especially if we are architects.

To go from the sublime to the ridiculous, I recently saw an example of environmentalist brutalism in a city not quite as famed as Paris for its beauty, namely Liverpool. Actually, Liverpool was once, at least in parts, a rather grand city, other parts of it being hideous beyond description, of course. It was once one of the largest *entrepôts* and passenger ports in the world; the proceeds of the slave trade in the eighteenth century had been invested in elegant Georgian buildings and the proceeds of the hugely expanded trade of the Victorian and Edwardian eras in grandiloquent municipal buildings.

A combination of economic change, the Luftwaffe, modernist bad taste, municipal corruption and union intransigence have reduced Liverpool to a kind of lived-in ruin, but it is not difficult to see a few remains of elegance and grandeur in it. I spent a few weeks in a part of the town where early Victorian stucco villas still stood, in a state of decay thanks to multiple occupancy, but not utterly beyond repair: the kind of houses that in more fashionable towns would cost a small fortune, though in Victorian times they had been built for only the lower reaches of the bourgeoisie.

One of the reasons that the area would never be renovated, however, was the council's insistence that each house had three large plastic trash-bins on wheels, each a bright colour: green for the bottles left over from last night's drunken orgy, red for stolen goods now surplus to requirements, purple for dead bodies and used syringes, all in fact that a modern British urban household needs to disembarrass itself of.

These bins completely dominated the appearance of the street; the eye simply could not avoid them. They were so ugly that to maintain the appearance of the houses would have been futile; the potentially elegant street would have remained an ugly dump, and therefore nobody did try to maintain the appearance of the houses. The justification for the bins, of course, was ecological: indeed, it was printed on them, like religious instruction.

Even very pretty villages in England are now ruined by this kind of environmentalist philistines.

Yet another example of environmentalist brutalism is, of course, the wind turbine. I am not informed enough to judge of their efficiency in terms of saving energy, let alone of their role in saving the planet, but I can certainly judge of their appearance. So could the late Senator Edward Kennedy who, though in favour of wind farms in general, was not in favour of wind farms anywhere he might have to look at them.

The minister in the last British government who was supposed to be in charge of the country's response to climate change (no problem, real or imagined, is without bureaucratic opportunity), and a man called Edward Milliband, said in a speech that it should now be regarded as socially unacceptable to oppose the establishment of wind turbines in one's own area. In other words, it should be deemed impermissible, a kind of social gaffe, even to question their efficiency, their noise, their effect on the beauty of the surroundings.

I suspect - though of course I cannot prove - that the real motive behind the espousal of wind turbines, at least in Europe, is quite other than the avowed one. It is not so much the need to save the planet as the desire to exert power (of financial corruption I will not speak).

In this respect I am reminded of Soviet-era iconography, particularly in its early phases, when drawn images of the countryside always included chimney-stacks in the distance belching black smoke. As we now know, the Soviets produced neither for profit nor use, but for power – their own, of course. The logic of the iconography was as follows: where there's black smoke from chimney-stacks there's a proletariat, where there's a proletariat there's a Party, and where there's a Party, there's no getting away from it, even in the remote countryside.

The purpose of the wind turbines is not, therefore, so much to produce electricity as to produce an effect, that of inescapability. You cannot look at a landscape with turbines, however distant, and not fix your eyes on them; they dominate, even from afar. Old ideologies might be dead, but environmentalism has taken their place with elegant ease. The power the turbines produce is more political than electric. They are the final triumph of the town over the country, of the metropolitan over the provincial, the utilitarian over the deontological.

This is not to say that the desire to minimise waste and pollution is not laudable. This is the aspect of environmentalism that I think is the most valuable. I confess, for example, that I find suburban gas-guzzlers offensive and as absurd as the Marlboro Man (who is about as genuine as Piltdown Man), but my reaction is as much cultural as ecological. I am appalled that, in a country such as Britain, and no doubt in all similar countries, a third of food is thrown away, either by supermarkets or individuals. No doubt less wastage might have a depressing effect on the GDP per capita: in which case, so much the worse for the GDP as a measure of the perfection of human existence.

Once again my objection to such waste is as much cultural as ecological: that people should carelessly waste that which for the vast majority of human history was a struggle to produce in even barely sufficient quantity seems to me a species of sacrilege. It implies a grossness in the approach life, a taken-for-grantedness of what should not be taken for granted, a lack of attention to what one is doing, that is not the way to a fulfilled human existence.

I am in favour of the refusal of supermarkets in Ireland and France now to hand out plastic bags, first because plastic bags are hideous in themselves, and second because they are indestructibly horrible once they escape into the environment. It is surely not too much to ask people to take sufficient thought to bring a bag with them when they shop; such thoughtfulness might (who knows?) even induce a more mindful attitude to what they are doing.

I am not, therefore, against actions taken to protect or preserve the environment that are above the level of individual conduct. It seems to me, though, that rather than indoctrinate children in the shibboleths of current ideological environmentalism, as seems to be happening in or schools, we should attempt to induct children into an attitude of mindfulness towards the world, in which they learn not to take all that is around them for granted in the pseudo-sophisticated manner in which they do so now.

For example, children should be taught to draw, preferably still-lifes. They would learn thereby to observe closely (military officers in the age before photography were taught to draw for intelligence-gathering purposes, and it is astonishing how many of them, who had presumably seen the rough side of life, ended up by producing tender drawings of the world for their own delight), and thus learn to appreciate the quieter pleasures that do not requite very much in the way of consumption. As the ancient author said, it is not in the fulfilment of desires, but in their limitation, that freedom resides. (Within reason, of course.)

Not every child will learn the implicit lesson taught by drawing, but this is hardly an objection. Most children will not change their conduct, their tastes or desires because of indoctrination with environmentalist shibboleths either, and those that do will probably become shrill ideologues of the kind for whom the price of certain current sacrifice is worth paying for the prospect of a distant utopia, and for whom the vision of that utopia (and its supposed apocalyptic alternative) is more real than the actual harm done in trying to bring utopia about that is before their eyes. Surely we had enough of such visionaries in the last century. The wish is mistaken for the fact: and so you ruin Paris and a thousands landscapes, but this is vastly less real to you than the marvellous distant prospect of renewable energy that shimmers in your eyes like a mirage in the desert.

There is another aspect to the hideous ecological house of Paris, the brightly-coloured waste bins of Liverpool and the wind turbines in idyllic rural settings: Man's (or at least certain men's) delight in destruction as a good in itself.

Mediocre but ambitious people – of whom, it seems to me, there are more than ever before in human history, the ambitiousness being what is new, not the mediocrity – are offended by the very sight of achievements that they know they cannot possibly match. They are not inspired by them, except to hatred and resentment. It offends them that the world should have achieved quite a lot before their advent into it; their response is therefore a destructive one. In Europe, architects rarely consider the harmony of what they build with what already exists, rather the reverse. That is why, in so many old towns in Europe, a harmonious assemblage of buildings, constructed perhaps over centuries, is comprehensively destroyed by one modernist structure. The hatred and resentment seeps out of it, like radon out of granite. Environmentalism is (or perhaps I should say can be) the new justification for the destructive urge born of resentment, all the more dangerous because of its plausibility.

21
Metsu-Metsu

E conomic crises come and go, but art goes on for ever; and so, while in Dublin recently, where there was what the late Saddam Hussein would no doubt have called the mother of all economic crises, I nevertheless went to an exhibition at the National Gallery of Ireland devoted to the work of the painter Gabriel Metsu.

Metsu, who died when he was only 37 years old, was once more famous and sought after than Vermeer, but their relative position in modern estimation has reversed. I am sufficiently a child of my time to agree with this reversal; but he was still a very fine artist indeed, and it is curious that there were no mobs of people clamouring to gain entry to the exhibition as there were at the last Vermeer blockbuster. Vermeer is a celebrity, Metsu is hardly known except to aficionados. For every book on Metsu, there is a library on Vermeer.

Though he started as a painter of historical and biblical scenes he is best known for his genre paintings of Dutch life of the Golden Age. He painted scenes from both ends of the social spectrum, and seemed as at home in a low tavern as in a high bourgeois household. Indeed, there was hardly any subject that he could not tackle; I wanted to take his still life with a dead chicken (belonging to the Prado) home with me.

Metsu painted several canvases of medical interest. In fact, he was as great a painter of the sick as Munch. In the Hermitage, for example, is a picture by him called *A Doctor's Visit*, in which a doctor, a sombre

121

bearded figure dressed in black, performs uroscopy - the supposed diagnosis of a patient by inspection of his or her urine - on the urine of a wealthy and richly-dressed woman sitting upright but ill in a chair. There is no hint here of satire, as there often is in genre painters' depictions of this bogus medical technique, as the doctor holds up the flask for examination; perhaps Metsu was not one of those who despised the elaborately-learned ignorance of the doctors of his day, who spoke an arcane language but could do nothing to help their patients.

Far greater than this painting, however, is *A Sick Woman and a Weeping Maidservant*, in the Staatliche Museum in Berlin. Again a rich but deathly pale and ill woman sits in chair, her eyes open but unseeing, her utter weakness so evident to us that we almost feel it in ourselves, a servant beside her, more or less in the dark, holding a cloth to her weeping eyes, her forehead contracted in grief. This is a deathbed scene, we know that all hope is vain, that the end is imminent and inescapable, that the woman has, like Faustus, but one bare hour to live.

And yet the picture is strangely uplifting rather than depressing. The maidservant's grief, depicted unmistakably as real, is testimony to the reality of human love and affection. Would there be love and affection among immortals, I wonder, where there was no possibility of separation between them? We could always postpone affection till tomorrow.

Another of Metsu's pictures is of a mother nursing a sick child on her lap. The listlessness of the child, the lack of tone in its limbs, is painted with incomparable, almost miraculous accuracy, which combines dispassion with compassion, precisely the combination that doctors must strive to achieve; and the greenish tinge of the child's complexion bespeaks a fatal outcome, though the tenderness in the mother's face is optimistic rather than despairing. By her is an earthenware jug with a spoon in it; one can imagine her trying to encourage the child to swallow something.

The tragedy of the picture is precisely the mother's apparent unawareness of the fatal outcome. Heartbreak is near, but she seems not to know it: that, ultimately, is the human condition (Metsu himself had only seven years to live when he painted the sick child). But in some situations heartbreak is nearer than others. Even in the Golden Age, when Holland was the richest, and probably the cleanest, country in the world, about a half of all children would not have survived into adulthood. The very prevalence of death makes the mother's capacity for denial all the more remarkable, and tragic.

There was another feature of the Metsu paintings that struck me immediately: the prevalence of dogs in them. Practically every portrait or social scene included at least one dog, a fact unremarked upon by the exhibition curators, who nevertheless lost no opportunity to point out the sexual symbolism in the paintings, indeed so insistently that one began to wonder why.

The dogs in the pictures were no mere mongrels, curs or flea-ridden pie-dogs, such as roam in scavenging packs in very poor countries. On the contrary, they were refined breeds, obviously well-fed and groomed. In most cases, it is true, the humans in the pictures take no notice of them, evince no particular regard for them; indeed, in many cases the dogs, by placing a front paw or paws on the dress of their mistress, while gazing up at them, seem to be begging for attention and love. This, too, is no doubt meant symbolically; but dogs that have been maltreated do not behave like this. They shrink from blows rather than seek embrace.

I suspect that the subjects of the pictures, once the scene is captured in paint, feel free to express their affection for their dogs: an affection that they deem is neither seemly nor dignified to display to the world, and have recorded for all to see forever. But the almost universal presence of these dogs - who are not working dogs - is enough to establish their importance in the life of that time. And no one would have gone to the trouble of looking after them who did not love them.

The relationship between man and dog is something that has interested me since the death of my own dog. How I (or rather we, my wife and I) loved him! We were never bored by, with or in his company. We had to control ourselves when people came to visit, in case they should see our extravagant love for him. 'Remember,' we would say to each other, 'we have to pretend for a time that he is only a dog. We can make it up to him afterwards.' We were ashamed, in case people thought we were psychological cases or emotional cripples, that we loved our dog so. As is well-known, some of the worst people in history have loved dogs.

We were not alone in our shame, of course; speaking confidentially, almost all people with dogs will confess to it. They too have to behave sometimes as if their dogs were mere animals. And how many people have I met who have said, after the death of a dog, that they could never have another because they could not tolerate the grief again!

When I worked as a prison doctor, I discovered that, when they first came into the prison, prisoners were frequently far more concerned for the welfare of their dogs than for that of the children they had fathered, generally very carelessly. Their affection for their dogs was deep and transparently genuine; they used their children as a pretext for access to their mother, who was invariably estranged from them.

Rationally speaking, I could not but find this appalling; but their love of their dogs nevertheless created a bond between us. By talking of their dogs to them, and letting them know that I, too, loved my dog, I established a kind of equality with them, not economic or social, of course, but - if I may be allowed a pretentious term - existential, and therefore deeper than other kinds of equality. We had something profoundly in common, that helped eliminate all other differences between us, even if the breeds of our dogs were very different. Prisoners, on the whole, had dogs of aggressive and even vicious type, however affectionate they were to their masters; but an English slum is not a suitable place for a lapdog, on the contrary. In an English slum, even a young man of diffident and pacific temperament needs to appear tough and dangerous. In this vile world, decency is construed as weakness, and weakness provokes attack. A man might want a poodle, but he had better appear with that kind of in-bred terrier whose jaws once locked into flesh cannot be prised apart except by decapitation: for the dog proclaims the man. Nevertheless, it is for their love in a loveless world that they are valued.

When I look back over the last years of my dog's life, I realise that I was both aware of his mortality - that, in the normal course of events, he must die well before me - and in denial of it. I never walked him without thinking of his death, of which (I assume) he had no conception himself. And yet, when he was nearly fifteen years old, and fell ill at last, I - a doctor! - managed to disguise from myself the inevitability of his impending death. I was like the mother in the painting by Metsu of the sick child. I averted my gaze from the obvious until the very last moment; no, until after the very last moment.

We took him to the vet for the nth time in n days. We were still looking for a cure. I was just like those patients who say to the doctor of their ninety-five year old parents, 'Surely there is something you can do.' By something they mean radical cure and rejuvenation, a recapturing of the past, a turning of the clock back.

When the vet who, I think, must have been building up to this in a gentle way, seeing how much we loved our dog, suggested that it was time to end his suffering, I suspected him of consulting his own convenience rather than the dog's welfare. It was nine at night; I suspected that he did not want to be called out much later at night or in the early hours of the morning. And yet I knew, of course, that he was right.

Up to the very last minute I hoped for a reprieve. I hoped that the vet would say he had suddenly thought of an alternative diagnosis, one that was indeed fully curable. When he injected a barbiturate intramuscularly to tranquillise our dog, prior to injecting him with a fatal dose, I hoped it would not work, that our special dog would somehow break the laws of pharmacology and not go to sleep. His little jump when injected - his last action in this world - is engraved painfully on my memory. How terrible that his last consciousness of the world (for I could not doubt that he had consciousness) should be of what to him must have been a painful, inexplicable and arbitrary assault.

And then I hoped that the vet would not be able to find a vein in which to inject him with the fatal substance. But he found one immediately, in a way in which, in another context, I would have found a manifestation of admirable skill, but which I now took as a sign of callousness. And then - last hope against hope - that the fatal substance would not prove fatal. All of this I hoped, and even after his death, when we laid him in his basket at home, I found it possible to hope that, despite the obvious cooling of his body, the supervening of rigor mortis, he would revive, jump up and demand his food.

The mother of Metsu's sick child deceives herself in the same way, as we must all deceive ourselves for much of our lives, if by no other means than by putting certain thoughts out of our mind. For, as La Rochefoucauld said, we can no more stare at death than at the sun. By a necessary aestheticising of the unbearable, Metsu protects our eyes.

22
A Version of Conversion

M any years ago, as a student, I read a paper in a medical journal that described four cases of religious conversion under the influence of temporal lobe epilepsy. At the time the paper delighted me, for I was not merely non-religious but anti-religious. It encouraged me to think of religious belief or experience as pathological in origin, though of course this was itself a pretty elementary error of logic, unworthy of the rationalist that I fondly supposed that I was.

The psychological, sociological or other origins of a belief tell us nothing about its truth. The fact that I believe that the Battle of Hastings took place in 1066 mainly because my teacher would rap me over the knuckles with her ruler if I did not answer her question about it with sufficient celerity does not mean that the Battle of Hastings did not take place in 1066 - or, for that matter, that it did. The Marxist notion that being determines consciousness and not consciousness being might serve as a rough and ready empirical generalisation, but cannot possibly be an invariant epistemological principle. The origin of many of our beliefs about the world, but by no means all or even most of them, is no doubt to be explained by our particular circumstances. So while I doubt that there is anyone who never uses *ad hominem* arguments, I equally doubt whether there is anyone who uses only *ad hominem* arguments.

I mention this as a preliminary to an examination of one of the most peculiar recent cases of religious conversion, that of Lauren Booth, half-sister-in-law to Mr Anthony Blair, former Prime Minister of Great Britain, to Islam. I write half-sister-in-law because she is half-sister to Cherie, Mr Blair's wife, and we live in a world in which the relationship of full sibling is becoming ever-rarer, like some insects or birds. The conversion of Lauren Booth, not surprisingly, made a few headlines.

In this case it is perfectly justified to resort to *ad hominem* remarks because Lauren Booth herself has advanced no other arguments or reasons than *ad hominem* ones for her own conversion.

She does not say, for example, that she became convinced that there was only one God and that Mohammed was his Prophet. On her own admission, she had read only a hundred pages of the Koran at the time of her conversion, so there was no possibility that she was persuaded by the sheer intellectual force of that document. To me, in any case, the Koran seems a terrible mish-mash (just as Carlyle described it); I fail to understand how anybody could be convinced by it intellectually. And only intellectual arguments call for replies that are not *ad hominem*.

Lauren Booth explained the reasons for her conversion to Islam in an article in *The Guardian* newspaper. There was no intellectual content to them whatsoever; the truth or otherwise of the doctrine did not exercise her mind at all, and was not even mentioned by her as a possible factor in her decision to convert. I accept that it is in the nature of conversion experiences that they should be more emotional than intellectual; but normally people are converted to what they believe to be a truth - thenceforth, for example, they know that their Redeemer liveth. For Lauren Booth truth (as either correspondence or coherence, or some combination of the two) seems to be no more important to than it was to her brother-in-law. She does not mention truth.

In so far as it is possible to extract reasons for her conversion from her women's magazine prose, here are the reasons she gives for converting to Islam:

> I began to wonder about the calmness exuded by so many of the "sisters" and "brothers."

> Second: The bending, kneeling and submission of Muslim prayers resound with words of peace and contentment.

> Third: Then came the pull: a sort of emotional ebb and flow that responds to the company of other Muslims with a heightened feeling of openness and warmth.

Fourth: How hard and callous non-Muslim friends and colleagues began to seem. Why can't we cry in public, hug one another more, say "I love you" to a new friend, without facing suspicion or ridicule?

Finally: I felt what Muslims feel when they are in true prayer: a bolt of sweet harmony, a shudder of joy in which I was grateful for everything I have and secure in the certainty that I need nothing more to be utterly content.

These are the kind of utterances of people who claim that they are spiritual but not religious; they are a cross between Dale Carnegie and relaxation therapy, with just a hint of eastern spice, but not too much. In fact it is distinctly more Californian than Iranian: Islam-tinged psychobabble. It is religion for those who think that multiculturalism is a matter of eating at a different restaurant every night. There is not a word about the truth or otherwise of the religious tenets of Islam.

Lauren Booth ends her article with a lesson on the deeper cultural meaning of 'Allahu akhbar!'

When Muslims on the BBC News are shown shouting "Allahu Akhbar!" at some clear, Middle Eastern sky, we Westerners have been trained to hear "We hate you in all your British sitting rooms..." In fact, what we Muslims are saying is "God is Great!", and we're taking comfort in our grief after non-Muslim nations have attacked our villages.

It would be breaking a butterfly on a wheel to elucidate all the lies and evasions in this short passage, but I do think that it has some significance from the purely *ad hominem* point of view, and that is her use of the first person plural and its possessive pronoun: we are taking comfort in our grief after non-Muslim nations have attacked our villages.

What we see here is the very characteristic modern thirst of people who have led privileged lives for the safe psychological haven of victim status (though I doubt that the thirst for real raw physical victimhood is not rather easily slaked). One of the attractions of Islam to someone like Lauren Booth is therefore quite likely to have been its tendency, which is no doubt cultural rather than doctrinal in origin, to self-pity.

Self-pity is one of the very few emotions that is sustainable for long periods, never lets you down, and is understandable by everyone: for surely no one so lacks compassion that he has not felt it at some time in his life. And self-pity oozes from the passage I have quoted like olive oil from fried aubergine.

In fact, there is another clue to Lauren Booth's conversion a couple of paragraphs before the end:

> In the past my attempts to give up alcohol have come to nothing; since my conversion I can't even imagine drinking again.

In other words, Islam is for her a kind of Alcoholics Anonymous, but with a bit more drama. If she had simply gone to AA in some church hall, there would have been no blaze of publicity; and she has been something of a publicity-seeker throughout her life.

The shallowness and egotism of her decision is further illustrated by how she continues, explaining how and why, thanks to Islam, she will never drink again:

> There is so much in Islam to learn and enjoy and admire; I'm overcome with the wonder of it.

One can almost imagine an advertising campaign based upon this: ISLAM, SO MUCH MORE TO ENJOY. TRY IT TODAY.

No doubt personal problems and a repudiation of conduct of the past play a large part in religious conversion, including that of St Augustine himself. And certainly there seems on Lauren Booth's own account to have been conduct to repudiate, as well as the bad example of her family. Her father, a minor comic actor who once played a role in a successful television comedy series, has had four wives and a life of alcoholic dissipation. By all accounts he has been neglectful of and abusive to those around him. His bad behaviour has been well-publicised and is known to all those who find the doings of minor celebrities of interest. He recently told the press that he did not love, and had never loved, his daughter Lauren, information that, true or not, would best have been kept private rather than bruited abroad. The fact that he saw fit to divulge it to a newspaper with a circulation of millions suggests what kind of man he is.

Lauren Booth admits to having drunk excessively and a quick scan of internet entries about her shows her to have been a full participant in the militant (and extremely stupid) vulgarity that is now the main feature of contemporary British culture, in which restraint has been altogether replaced by the crudest possible frankness. It is well that she should turn her back on her own past. A single mother of two who drinks too much and behaves crudely in public is not much of an example. With advancing age, of course, suggestive comportment attracts less amusement than pity and disgust, even among the most vulgar and dissipated of onlookers.

But why Islam? It is not as if other religions, above all Christianity, have not offered people with chequered pasts the occasion and opportunity to reform and to eschew past indiscretions. And since we have seen that the question of religious truth does not even occur to her, it is to her psychology that we must look for an answer.

I suspect that something similar to the psychology that induces black prisoners to convert to Islam has operated on her, but with an extra twist in her case.

Black prisoners, like others, want eventually to give up their life of crime. It is clear from statistics that most criminals give up by, around or at the age of forty. Perhaps this has something to do with declining hormonal drive, or with maturing mentality. Be that as it may, many criminals seek or require an explanation of their own changed inclinations, and religious conversion is such an explanation. That is to say, religious conversion is as much an expression as a cause of a desire for change.

However, there is a lot of cultural pressure on people nowadays to appear not to conform to what were once the mores of society, especially among those who come from discontented or resentful sectors of society. Individuality, the most desired but elusive of attainments in mass society, seems to require the repudiation of what was once deemed respectability. It is not respectable to be respectable. And you would have to be very tone-deaf to the music of western society not to realise that conversion to Islam is a good way to alarm it, not to be respectable.

In the case of Lauren Booth, there was the additional incentive of publicity. If she had converted to evangelical Christianity, and even had started speaking in tongues, no one would have noticed. It would hardly have merited even a line in a local newspaper. But she had only to don a pink hijab (which, it must be admitted, is more becoming than anything else she has ever worn) for a flurry of publicity to follow. And, one suspects, she belongs to the ever-enlarging category of people to whom to be is to be seen; to be out of sight is not merely to be out of mind, but to cease to exist, even for oneself. She is certainly not the only member of her family to suffer from this most unfortunate and morally debilitating of conditions.

There is, of course, one further motive that might have impelled her. Her conversion was almost certain to give a momentary frisson of embarrassment to her half-in-laws, Mrs and Mrs Blair, whose kleptocratic sanctimony might very well have irritated her, principally because of its apparent and continuing success. She had few cards in her hand to play; her attempts at public vulgarity had failed to distinguish her in any way from the general run of vulgar public figures, or seriously to embarrass her half-in-laws. Conversion to Islam was a trump card, and she has now played it.

23
Beauty and the Best

A controversy recently erupted in Sweden over an article published by the philosopher, Roger Scruton, in a magazine called *Axess*. He argued in it that Western art no longer had any spiritual, let alone religious, content; indeed, it had become afraid of the beautiful, from which it shied away as a horse from a hurdle too high for it. The result was a terrible impoverishment of our art.

The same magazine had published, shortly before, an article about Islamic art in which the author said that such art was inseparable from the religious ideas and beliefs that it embodied. This passed without remark: no one wrote in angrily to say, 'So much the worse for Islamic art.'

Professor Scruton's suggestion that western art had become impoverished as a result of its radical repudiation of anything transcendent in human existence in favour of the fleeting present moment, however, exasperated and infuriated the professional art critics of Sweden – as, indeed, it would have done the art critics of any western country. They reacted with the fury of the justly accused: for it is the professional caste of cognoscenti who have consistently applauded the trivialisation of art and its relegation to the status of financial speculation at best, and a game for children showing off to the adults at worst.

There was a good example both of art as financial speculation and as silly game at Versailles recently, where some of Jeff Koons' sculptures were shown in an exhibition. I am no great lover of Versailles myself: it strikes me as pompous and overblown, and its formal perfection does not make up for this. Still, no one can fail to recognise its magnificence, and its peculiar unsuitedness to the display of Koons' cheap and childish artefacts (I mean cheap in the moral, not the financial, sense, of course). It is impossible at Versailles, even if one is no egalitarian, not to think of the immense exploitation of the peasantry upon which it was raised; to exhibit Koons there, whose work is a knowing joke (and not a particularly good one) repeated over and over again, is a final insult to the memory of those whose years of toil made Versailles possible. The least that is due to their memory is to use Versailles for something worthwhile.

The mere exhibition of the work at a location such as Versailles serves to keep its (monetary) value up, to save those foolish enough to have invested in it the embarrassment not only of being shown to have no taste, but – worse still in the circumstances - no financial acumen either. In no field is Hans Christian Andersen's fable about the Emperor's new clothes more salient than contemporary art; or, to put it another way, in no commercial field are there so many Bernie Madoffs.

But, you might say, Jeff Koons is an original. No one had ever done anything like his sculptures before, and after two millennia of artistic endeavour it is no mean feat to be original. You can recognise his work anywhere; indeed, it is quite unmistakable. So, of course, is the logo of Coca-Cola or Hitler's face: unmistakability is not by itself a criterion by which art ought to be judged. The chief interest of his work is sociological, not artistic: how can it ever have been thought worthwhile?

It is here that Scruton's argument becomes illuminating. The successful modern artist's subject is himself, not in any genuinely self-examining way that would tell us something about the human condition, but as an ego to distinguish himself from other egos, as distinctly and noisily as he can. Like Oscar Wilde at the New York customs, he has nothing to declare but his genius: which, if he is lucky, will lead to fame and fortune. Of all the artistic disciplines nowadays, self-advertisement is by far the most important.

This is reflected in the training that art students now undergo. Rarely do they receive any formal training in (say) drawing or painting Indeed, from having talked to quite a number of art students, it seems that art school these days resembles a kindergarten for young adults, where play is more important than work. The lack of technical training is painfully obvious at the shows the students put on. Many of the students have good ideas, but cannot execute them successfully for lack of technical facility. Indeed, their technical incompetence is only too painfully obvious.

It is very striking, too, how few art students have any interest in or knowledge of the art of the past. Do you visit galleries, I ask them?

No, they reply, a little shocked at the very suggestion, and as if to do so would inhibit them in their creativity or to condone plagiarism.

As for art history, they are taught and know very little. This is all part of the programme of disconnecting them radically from the past, of making them free-floating molecules in the vast vacuum of art.

It is true that they are sometimes taught just a little art history. I had what was for me a memorable conversation with an art student when she was my patient. She was in her second year of art school, and told me that one of the things she enjoyed most about it was art history. I asked what they taught in art history.

'The first year,' she said, 'we did African art. But now in the second year we're doing western art.'

I asked what particular aspect of western art they were doing.

'Roy Lichtenstein.'

As satire would be impossible, so commentary would be superfluous. The task is not so much to criticise as to understand: that is to say, to understand how and why this terrible shallowness has triumphed so completely almost everywhere in the west.

No such question can be answered definitively; but I would like to draw attention to two errors that have contributed to the triumph of shallowness. The first is the overestimation of originality as an artistic virtue in itself; and the second is the false analogy that is often drawn between art and science in point of progress.

Let me take the second point first. One often hears of 'cutting-edge' art; indeed, the much older term, *avant garde*, is of the same ilk. This suggests that there is progress in the arts, as there is in science, and that what comes after must, in some sense, be better than what came before. Art has some kind of destination, with later artists further along the road to it than earlier.

In science, progress is a fact (except for the most extreme of epistemological sceptics, none of whom, nevertheless, would be entirely indifferent as to whether their surgeon used the surgical techniques of, say, the 1830s, rather than those of this century). The most mediocre bacteriologist alive today knows incomparably more that did Louis Pasteur or Robert Koch, for example; the most mediocre physics graduate knows incomparably more than Sir Isaac Newton. This is because scientific knowledge is cumulative. But no one would suggest that the paintings of Rothko were better than those, say, of Chardin because he lived a long time after Chardin, and that Chardin's were better than those of Velasquez for the same reason.

Art teachers and critics use the false analogy with science in order to deny the importance of tradition in artistic production. They do not realise that science is entirely dependent on tradition for its progress. It is not just that most competent scientists know a lot about the history of their subject, but that the very problems that they set about solving, their entire mental worlds, are inherited by them. No scientist has to discover everything anew for himself: no mind, however great, is expected to begin again from zero. Tradition is the precondition of progress, not its antithesis or enemy.

Since art makes no progress, the role of tradition in it is obviously very different. But it is no more realistic to expect artists to be able to fashion *ex nihilo* a worthwhile response to the world and their own experience of it than it is to expect every schoolboy to discover Newton's laws of motion for himself. Genius is supposed by modern theorists of art education to be like flies according to the spontaneous generationists of the pre-Pasteurian age, who thought that they emerged from decomposing matter by a kind of spontaneous alchemy, rather than from eggs laid by older flies. Worthwhile art grows out of other art: it neither repeats it (that would be impossible even if desired or desirable), nor is totally ignorant of it.

There is a paradox in the new ideology of art with regard to originality. It is both overvalued, and at the same time believed to be too easily achieved.

In a sense, everything that human beings do is original, for even if they want to they cannot exactly copy one another. Like M. Jourdain speaking prose, most of us utter completely original sentences by the dozen every day, effortlessly and not knowing that we are doing it.

This is not the kind of originality that is valued in the new art ideology. What the new art ideology means by originality is that which has the power to shock, especially the bourgeoisie (if it still existed). Only the rebellious is original and creative: Norman Mailer, for example, in his essay *The White Negro*, equates rebelliousness and creativity, by contrast with 'slow death by conformity.'

Unfortunately, deliberately setting out not to conform results in a conformity of its own, one indeed that has become a mass-phenomenon.

Non-conformity for its own sake cannot be the source of true or valuable originality, therefore. The only kind of non-conformity that leads to worthwhile originality is the unselfconscious kind, that arises because the person has something new to express that is transcendently worthwhile, *sub specie aeternitatis* as it were, and that might or might not lead him into conflict with others. Doctor Johnson, with his usual penetration, has it absolutely right:

> Singularity, as it implies a contempt of the general practice, is
> a kind of defiance which justly provokes hostility or ridicule;
> he, therefore, who indulges in peculiar traits is worse than
> others, if he be not better.

Originality is not, therefore, a virtue in itself, moral or artistic; and a man who sets out to be original without both the technical ability to express something new, and (most important of all) the possession of something worthwhile new to express, is merely egotistical. That is why the art critics, who are inclined to praise works as being original, path-breaking, taboo-breaking and transgressive, without any reference to their transcendent worth, are wrong, and Roger Scruton is right.

Where does the fear in modern art of such qualities as beauty and tenderness towards the world come from? (I am talking here of art that achieves public notice and notoriety: there may be hundreds or thousands of excellent artists who fear neither beauty nor tenderness, but whose work goes unremarked.)

I think it has something to do also with our inflamed egotism, that requires that we should be entirely self-sufficient and autonomous, philosophically, morally, intellectually and economically.

Beauty is a fragile and vulnerable quality, and moreover one that is difficult to achieve; ugliness, by contrast, is unbreakable and invulnerable, and very easy to achieve. (How easy it is to look bad, how difficult to look good!) By espousing the ugly, we make ourselves invulnerable too; for when we espouse the ugly, we are telling others that 'You can't shock, depress, intimidate, blackmail, or browbeat me.'

We use the ugly as a kind of armour-plating, to establish our complete autonomy in the world; for he who says that 'I find this beautiful,' or 'This moves me deeply,' reveals something very important about himself that makes him vulnerable to others. Do we ever feel more contempt than for someone who finds something beautiful, or is deeply moved by, what we find banal, trivial or in bad taste? Best, then, to keep silent about beauty: then no one can mock or deride us for our weakness, and our ego remains unbruised. And in the modern world, ego is all.

24
Evil Be Thou My Evil

O ften I read more than one book at a time. When I tire of one I fly to another. This is because the world has always seemed to me so various and so interesting in all its aspects that I have not been able to confine my mind to a single subject or object for very long; therefore I am not, never have been, and never will be the scholar of anything. My mind is magpie-like, attracted by what shines for a moment; I try to persuade myself that this quality of superficiality has its compensations, in breadth of interest, for example.

Be that as it may, I recently spent a day reading two books and constantly switching between them; the first *Mao's Secret Famine*, by Frank Dikotter, a professor of Chinese at the London School of Oriental and African Studies, and the second *Beyond Evil*, by Nathan Yates, a journalist on the British tabloid newspaper, the *Daily Mirror*.

The famine of Dokotter's title was that brought about by the Great Leap Forward in China between 1959 and 1962; the thing which was beyond evil was the murder of two little girls in the Cambridgeshire village of Soham, whose disappearance for a time captured the attention of the world.

The famine was probably the worst in world history, at least as measured by the absolute number of victims; according to Dikotter, there were 45,000,000 of them. Other famines have been worse relative to the total population: it is sobering to recall, for example, that the population of Ireland is still only 70 per cent of what it was before the great potato famine of the 1840s, when some in Britain regarded it coldheartedly as a Malthusian winnowing of the surplus population ordained by God.

Nevertheless, there seems something peculiarly dreadful about the famine caused by the Great Leap Forward: from its complete predictability from previous human experience of such great leaps to the utter indifference of Mao Tse-tung to the deaths of scores of millions of his compatriots and to the suffering of hundreds of millions of more of them.

Of course, Mao could not have produced the famine single-handedly, but the other authors of the catastrophe acted mainly from cowardice or sycophancy, unattractive but nevertheless human qualities that few of us have never exhibited in the course of our lives.

One finishes Dikotter's book with a visceral loathing of Mao, whose preparedness to contemplate seriously the deaths of hundreds of millions of people if only his dim and half-baked ideas about the good society might be put into practice places him among the select company of true moral monsters of the Twentieth Century, for example Lenin and Hitler. The only lesson that Mao drew from the Great Leap Forward and its terrible associated famine was that he should revenge himself on Liu Shao-chi, who did much to bring it to an end. Mao duly took his revenge on Liu during the Cultural Revolution, again at the cost of untold suffering and destruction. Mao cared about as much for humanity as most of us do for ants in the kitchen.

By comparison with the deaths of 45,000,000 people, those of the two little girls in Soham might seem insignificant. What are two to set against so many? The culprit was a man called Ian Huntley, who had a long history of dubious sexual relations, including with 12 year-old girls. But though he had been reported to or investigated by the police several times (many witnesses, including the mother of a 12 year-old girl whom he had sexually assaulted, refused to testify against him), he had no convictions and therefore no criminal record; thus, when he applied for a job as school caretaker there was no official record of anything that he had done that precluded him from taking up the job. What amounts, legally-speaking, to tittle-tattle cannot be allowed to stand in a man's way.

The two girls whom he murdered went for a walk one evening, and he asked them into his house. They suspected no ill of him because they already knew him from their school. What happened next is known only to him, who has never told anyone the truth of what he did. Probably he assaulted one or both of them sexually and then, realising that each would be a witness to any allegations made by the other, he killed them both, taking their bodies under cover of night to a distant ditch where he later burnt them beyond recognition.

When a hue and cry for the missing girls was raised he played to psychopathic perfection the part of a concerned person and good neighbour. With all the other villagers, he searched high and low for the two girls, though of course he knew all along where they were. He even uttered words of comfort to the distressed parents.

The question I asked myself as I read the two books, switching from one to another, is 'How and on what scale do you compare the evil of the two men, Mao Tse-tung and Ian Huntley?' It hardly seems satisfactory to say that Mao was 22.5 million times worse than Huntley because he was responsible for that many more deaths than he. And yet to utter the two names in the same breath seems almost to indulge in bathos.

Huntley probably did not set out to kill the two girls. He had been violent to women and adolescent girls before, but not so as to cause them permanent physical injury. The chances are that in killing them he was only trying to get rid of the evidence. He preferred their deaths to his exposure as a sex criminal.

Likewise, Mao did not set out to kill millions, but he much preferred to do so than have to back-pedal or re-think his ideas, a back-pedalling that would have cost him his power. If the implementation of his ideas led to disaster, therefore, it proved to him only that there were saboteurs, class traitors and capitalist-roaders still at large, enemies to be destroyed. Never could he admit that his ideas were wrong, and moreover wrong for obvious reasons that anyone of average intelligence ought to have been able to see. Let the heavens fall, he said to himself, so long as I preserve my power.

It is clear that the evil of these two men cannot be compared using a linear scale, and the same goes for the suffering of their victims. Who would expect the parents of the murdered girls to be consoled by the thought that at least the murder of their children was not the Great Leap Forward, that at least Huntley killed only two to Mao's millions, and that, everyone else they knew apart from the girls had survived, which was certainly not the case during the Great Leap Forward, when every survivor knew of scores who had died? And what would one think of a defence lawyer who argued in court that Huntley was not as bad as many others in history that he could name if he wanted?

Huntley and Mao did what evil they could within their own

spheres. Mao's sphere, alas, was the largest population in the world, while Huntley was confined to a small village in England. Only one of them – Mao – got away with it. But both conscientiously did the worst they could.

The urge or temptation to place people in a league table of evil is very strong, as if evil were measureable on a linear scale, like height or weight. Was Stalin as bad as Hitler, and if not, by what percentage was he less bad? Twenty per cent, forty per cent? Serious arguments are held on this question; I have had them myself, as if something depended upon the answer, as if indeed there were an answer; likewise the comparison of communism with Nazism.

Protagonists of the view that communism was at least as bad as Nazism point to the fact that it killed more people. Protagonists of the opposite view say that, while this might be so, communism lasted seventy years, while Nazism lasted only twelve, as if a longer rule implied almost a right, or an excuse, to kill more victims. If one divides the number of victims by the number of years in power, Nazism was probably worse than communism, even if it is not always entirely clear who was the victim of which ideology. Therefore, goes the argument, Nazism was the worse.

Then, of course, there is the argument about intentions. Communism may have been responsible for more deaths than Nazism, but at least it killed in the name of a universal ideal, not in pursuit of the supposed benefit of only a small portion of mankind. This is a distinction that has always seemed to me rather odd. Who would be much consoled by being asked whether he would prefer to be brutally murdered in the name of a universal ideal or merely because he was a member of a hated racial or religious group? Is it better to be killed as a bourgeois, as a kulak or as a Jew?

In opposing evil, of course, we often commit acts that, in other contexts, would themselves be evil. We are tempted to suppose that the end justifies the means – which sometimes it must, of course, but not always, if for no other reason than that the connection between ends and means is inherently an uncertain one.

And there is another trap that awaits us: there is nothing more delightful to the human mind, or at least to many human minds, than to do evil in the name of good. The number of sadists is legion, and the impulse grows with its satisfaction. Even the dullest of understandings is lightning-quick in its capacity to rationalise sadistic urges in the language of morality. We can thereby come easily to resemble, even if in only attenuated form, those whom we so fiercely oppose and claim to abhor.

There was a remarkable instance of this in the book about the Soham murderer, Ian Huntley. When he was brought to court for his trial, a mob had gathered outside to hurl execration at him. Most in the mob were women, and many had their young children with them. These children screamed in terror as their mothers threatened the culprit (still technically innocent) with physical violence. There is little doubt that, had it not been for the presence of the police, the accused would have been torn limb from limb, children or no children.

But it is morally certain that these Mesdames Defarge lived lives that were not beyond reproach as far as their upbringing of their children was concerned. At the very least, the example of public behaviour that they set was appalling, and their disregard of the terror of their own children in itself a form of abuse. Almost certainly some of them were the kind of women who would have refused to cooperate with the police in the days before Huntley turned murderer. If reproached for their behaviour, they would, with that quickness of mind to which I have already referred, have returned the reproach to its sender by saying that he who made it was a sympathiser with Huntley, and therefore some kind of accomplice of his.

The mob that howled at Huntley resembled that which howled at the victims of the Cultural Revolution. It is true that, unlike the victims of the latter, Huntley had committed a real and terrible evil, but it was an evil so self-evident that it required no howling mob to make evident or condemn it; those who suffered most from it were certainly not among the mob baying for (among other things) his death.

Delight in evil is very widespread, even if it is not quite universal, and it takes many forms. I do not exclude myself from these strictures, for I have sometimes enjoyed inflicting suffering on others, even if only of a comparatively mild kind. I have even sometimes suspected that I have enjoyed living among, and reading about, evil in order to assure myself that I am a jolly good fellow, at least comparatively-speaking, using a linear measure of evil of course.

25
Anyone for Bunga-Bunga?

I t is very wrong of me, no doubt, but I have been rather enjoying the Berlusconi sex-scandal. Of course, by British standards it is all rather tame, being merely a matter of orgies with scores of nubile young women; we prefer our politicians or prominent people to be flogged by a dominatrix dressed as a concentration camp guard, or as a very minimum to indulge in autoerotic asphyxia. Poor old Silvio seems sadly lacking in imagination.

I must admit that I find it rather difficult (and not altogether pleasant) to imagine him in any kind of sexual activity whatever. Recent photographs of him make him look like something out of Madame Tussaud's, most likely an escapee from the Chamber of Horrors, with his implanted hair and waxy visage. He could have a new career playing Dracula.

Naturally, the thing that really upsets him about the current allegations, apart from the fact that they might land him in prison, is the suggestion that he paid for sex. Supermen do not pay for sex, they seduce women who queue up to be seduced. But if Berlusconi didn't pay for sex, it must be about the only thing that he hasn't paid for.

Latin countries are not supposed to have sex scandals, they are too sophisticated or relaxed (not quite the same thing) about matters sexual for that. I have some sympathy with this view, because the prurience of the Anglo-Saxons is both ridiculous and generally-speaking hypocritical. It is also boring, like feminism, critical theory and baseball. Besides, if it is eroticism that you want, few subjects are less arousing than, say, an account of Mr Clinton's sex life. His antics were surely enough to make an erotomaniac blush and turn over a new leaf.

There are interesting aspects of the Berlusconi case. He is accused of prostituting minors, though I don't think (to judge from the photographs of her) that the person principally involved, whom I won't name out of respect for her supposed minority, would get a part as Little Red Riding-Hood: Mata-Hari, maybe, or the role of Jane Russell in a film of that film star's life, but not Little Red Riding Hood.

The age of legal consent to sexual relations in Italy is sixteen. Those who are younger than that may indulge in sexual activity, but only with those of similar age to themselves. This, of course, is to prevent the abuse of authority that older people have over younger.

I have seen no suggestion that Berlusconi has had sexual relations with girls under the legal age of consent in Italy. And presumably if one is deemed old enough to consent to sexual relations at the age of sixteen, one is deemed old enough to choose with whom one is to have them, and (within limits, naturally, that apply to everyone) under what conditions.

When Berlusconi says, then, that the interest of the legal authorities in his sexual conduct (which he denies) is part of a political plot against him, his allegations are not to be dismissed out of hand. When one hates someone politically enough, one is tempted to use any instrument that comes to hand to destroy him, legitimate or not. This is so whatever side of the political fence (assuming there to be two, rather than several) one is on.

We come now to the more difficult part of the question. Let us suppose for the sake of argument that Berlusconi has done nothing wrong legally in consorting with girls of 16. The law has nothing to reproach him with, therefore: does that mean that we, or the Italian electorate, have nothing to reproach him with?

It seems that the Italian electorate, or at least a large part of it, admires Berlusconi precisely because he cocks a snook, if not at authority precisely (as Prime minister he is, after all, an important authority himself), at least at orthodoxy. He has escaped into a realm where he does not have to obey anyone or care what anyone thinks of what he says and does. And who among us has never dreamt of being in that situation? Berlusconi lives out the fantasies of millions of his compatriots; they think that he is the only really free man in Italy, and as he is self-made, they can dream of being like him.

Only he could have had the effrontery to imply that the reduction of youth unemployment has been the first concern of his life, and leave it to the public to conclude that his orgies have been arranged with this in mind: a small contribution, no doubt, to the solution of the problem but a contribution nonetheless. The fact that this insinuation is incompatible with the denial that he ever paid for sex is not very important.

When an earthquake devastated a town in central Italy, making thousands homeless, Berlusconi told them to think of it as a camping holiday. From the point of view of the homeless townsfolk, living among the ruins, this must have appeared as what it was, deeply callous; from the point of view of at least many Italians, however, it must have come as a relief that they did not have a leader who claimed to feel more than he actually did feel, or to feel everyone's pain as if it were his own. In its way, emotional inflation is as bad as other kinds. If we must have leaders who are psychopaths (and it seems that we must), let us at least have those with the courage of their indifference. Better a Berlusconi than a Blair, with all his fake, Clintonesque emotion. (Blair was Clinton, but without the self-awareness.)

Of course, to say the opposite of what everyone else would say in the circumstances is not an infallible method of arriving at the truth, or even of something worthwhile to say; and by no means all conventional expression is despicable. We console the widow, even though we detested the spouse, and this is as it should be. To refuse to say what everyone else would say is more often a sign of egotism than of original thought; but such is the crushing and suffocating nature of current emotional correctness that Berlusconi's gaffes come as a relief: for we are all accustomed now to claiming to feel more than, in our hearts, we do feel. One sometimes thinks that if Berlusconi did not exist, it would be necessary to invent him.

To return to the question of Berlusconi's conduct. To exonerate him on the grounds that his conduct was within the letter of the law is, in my opinion, to give the law too much moral importance. It does not follow from the fact that such-and-such a course of conduct is within the law that it is permissible in any other sense. No one thinks that, because insults in public between man and wife are not illegal, they are therefore morally acceptable or neutral.

However, on the view that all human relationships, at least between adults, are contractual in nature, Berlusconi (assuming the stories of orgies are true) has done nothing with which to be reproached.

Rightly or wrongly, the law deems the participants in these orgies to have been adults perfectly mature and able enough to make their own decisions, which includes the wrong ones; if orgies with an old man of what Shakespeare calls a 'goatish disposition' is what they want, with or without payment, then it is no one's place to complain. Only coerced orgies are wrong.

Now it might be said that the difference in age and authority and social position and financial situation between the parties in these orgies – on the one hand, a 74 year-old Prime Minister who is also one of the richest men in the world, and on the other a group of young women (I dare not call them girls), many of them of disfavoured backgrounds and none of them rich – was so great that it was a grossly unequal contract; and grossly or unreasonably unequal contracts in law are not enforceable.

But either people are free to make their contracts or they are not. No doubt there is a continuum between contracts that are so unequal as to be unfair and those that are between fully equal parties; there is no sharp dividing line between the ends of a spectrum, but the law needs to draw such a line.

The need for boundaries that are to some extent arbitrary, and which cannot fully capture social realities, was brought home to me when I discussed the question of underage sex with British parents. The law in Britain prohibits sexual relations before the age of sixteen; and although parents can be imprisoned for failing to send their children to school, or even merely failing to get them there against their will, until the children are sixteen, doctors are prohibited from informing parents that they have prescribed contraceptives for their children, even of twelve or thirteen. Many parents, also, connive at their children's sexual relations at an age well before that of legal consent. All this seems to me to bespeak a certain cultural confusion on the matter of when children are, or should be, free to make their own decisions.

When I discussed the matter with parents (or, more usually, a parent, the other parent having abandoned his responsibilities a long time before), one argument was invariably used by the parents: that it is absurd to suppose that a child who is 16 years old can make a decision that a child who is 15 years and 364 days old cannot. Never did a parent argue that the age of 16 was wrong and unrealistic, and that it should be revised downwards, to say 14 (it is 13 in Spain).

The problem with their argument, of course, is that it dissolves all possible age boundaries whatever; for no age will be found, immediately before which a person is mature enough to take important decisions, and immediately after which he or she is sufficiently mature. Therefore, if there are to be boundaries at all, and most people think that there should be, they must to an extent be arbitrary, even if they must bear some relation to a reality, and not just those that each person thinks up for himself to suit himself as an occasion arises. And the boundaries work in both directions: what was impermissible before becomes permissible after.

This means that, under the rule of law, we have to accept certain conduct that we find distasteful or even disgusting, such as that alleged of Berlusconi. But because we have to accept it in the legal sense does not mean that we have to accept it in any other.

There is the further question of whether his private life (which seems to have been pretty well documented for a private life) is anybody's business but his own. Even though people project themselves into the public sphere, they still have a right to a private life: they do not become public property, nor have we the right to demand of them that they should be exemplars of virtue in all respects.

On the other hand, we have a right to demand of them that, in return for entrusting them with high office, they behave with reasonable discretion and dignity. Whatever they do, whatever vice they indulge in, it should not be front of the children, namely us. This means that we must accept hypocrisy as being more desirable than consistency. But games of bunga-bunga in the basement of the villa of San Martino in Arcore are not really compatible with high office; and Berlusconi is enough of a man of the world, and of the media, to know that such activities are unlikely to remain hidden from public knowledge forever. It is true that he controls much of the press and the broadcast media of Italy; but by no means all. Recent events in Tunisia show that no one can or does control all the media of information, even of countries much less pluralist than Italy.

In reading of Berlusconi's alleged antics (which, of course, I assumed to be more than merely alleged), I became aware of an unexpected emotion: pity for him, or sorrow. He seems to be a man whose absolute freedom requires that he should accept no limits. As Shigalov says in *The Devils*, 'Starting with absolute freedom, I end with absolute tyranny.' In Berlusconi's case, it is the tyranny of his own limitlessness.

26
Of Termites & Mad Dictators

T o hate a tyrant is not to love liberty: rather more is required than that. And with liberty as with all other objects of affection, the course of true love never did run smooth. It is unlikely to do so in the Middle East.

Of all the tyrants, Muammar Ghaddafi was undoubtedly the worst. If he were not so sanguinary, if he had not brought permanent civil war to so many parts of Africa, if he had not ruled so consistently by terror, he would have been a figure of fun, more to be derided than hated: a preposterous semi-lunatic with bad taste let loose in the store of a theatrical costumier, who thought himself valiant by pinning a made-up medal to the chest of his own made-up uniform, and who frequently dressed as if he had asked Armani to design a costume incorporating the Bedouin and Ruritanian traditions, with just a hint of African tribal leader thrown in. Of course, it is not so very long ago that he had his admirers among the left-leaning intelligentsia of Europe and elsewhere, who bent over backwards to understand (which is to say, excuse or deny) his dictatorial propensities as the natural result of Libyan history. But, apart from those whom he has bribed and Anthony Blair, former Prime Minister of Great Britain, he has few admirers now.

Ghaddafi was so loathsome a figure, in fact, that (much to my self-disgust) I found myself in my imagination suspending my objection to lynching in his case. It would seem no more than poetic justice that he should meet the same fate as Mussolini, whom he in so many respects resembled. It was only with difficulty that I managed to suppress this primitive reaction, and remind myself that to meet brutality with brutality, to match savagery with savagery, is seldom the way to a more civilised future.

Some of the auguries in the Middle East are not entirely favourable. There were cries directed at Ghaddafi in Libya of 'Go back to Israel:' a reference to the rumour that his mother was Jewish, and not exactly indicative of that broadness of mind that is necessary for the establishment of a free political system. As to Egypt, polls have indicated in the recent past a large majority in favour of the death penalty for apostates, again not a good principle on which to found a free country. 'Life, liberty and the pursuit of heretics' doesn't have quite the right ring to it.

Not should it be forgotten that hatred of being oppressed is not quite the same as hatred of oppression itself. There have been enough instances in history of the oppressed turned oppressor for this hardly to need emphasis. And this is so especially where the oppressed believe themselves to be the bearers of true doctrine as against the bearers of the false doctrine that has hitherto oppressed them. What is the point of being free unless you can impose yourself on others, and make the true doctrine rule the world?

There is a very simple problem in the Middle East: simple, that is, conceptually, not simple from the point of view of finding a practical solution to it. Islam has not found a doctrinal way of rendering unto Caesar those things which are Caesar's; and since one of its founding principles is the inequality of man so long as not all men are Muslim, equality before the law is very difficult to establish in a country with a preponderance of Islamic sentiment. Either it must be imposed by a secularising elite, in which case it is felt as oppressive and anti-democratic, in the sense of being against the wishes and feelings of the majority; or it simply fails to exist. And where it does not exist, modernisation can be but a veneer.

At best, then, equality before the law – an essential condition of the rule of law – is precarious and likely to be more honoured in the breach than in the accomplishment. Not even in countries that invented the rule of law is it ever flawlessly adhered to; but in countries where it is imposed, it is likely to be but an ideological fig-leaf for a tiny kleptocratic elite; an elite that comes to be so justifiably hated that it in itself provides an *ad hominem* argument for those who argue for a return to the supposed (but historically false) prelapsarian purity of Islam. The choice, then, is between Scylla and Charybdis; it will not be easy, short of a loss of Islamic faith by a multitude of people, to escape this unappealing dilemma.

It might be argued, of course, that as yet the revolts in the Middle East have had little by way of specifically Islamic content. This is true; to judge by the way they dress, the youthful demonstrators could have emerged from any working class area in any European or North American city. Their demands have been democracy, liberty of expression and so forth. But liberty of expression is a two-edged sword where a large part of the population is viscerally opposed to it.

It would surely be a mistake to assume that the crowds in Tahrir Square, impressive as they undoubtedly were, represented faithfully the entire population of Egypt, which is so many times larger. Whatever government emerges from the current uncertainty, it will have to face the potentially enormous and destructive gulf between the *pays réel* and the *pays légal*, if the *pays légal* adopts the rule of law as its founding principle. And even if the leader of such a government should turn out to be personally incorruptible, it is vanishingly unlikely that the immense apparatus beneath him should be likewise incorruptible. Thus the whole sorry vicious circle will be set up again, and the problem unresolved.

Even if Islam were to lose its grip on the population, the outlook might not be rosy: for when the people lose their faith in Islam, they will not believe in nothing, they will believe in anything. And among the things they might come to believe in is some kind of secular salvationist doctrine which – if the history of the Twentieth Century is anything to go by, and it is the only guide that we have – might be even more sanguinary than what they previously believed in.

As for the future of liberty in our own countries, in our own civilisation, we have no grounds for complacency or boundless optimism. In Europe, for example, a gulf that grows ever-wider has opened up between the *pays réel* and the *pays légal*, between the sensibility of the self-replicating politico-cultural complex (to adapt slightly Eisenhower's famous phrase) and the sensibility of vast numbers of the population. The *pays légal* has imposed, among other things, a unified currency on the continent against the wishes of the population in so far as they were ascertained, which is no small thing to have done when one considers the political and economic tensions to which that unification was always bound before long to give rise. The more detached from the *pays réel* the *pays légal* becomes, the greater the chance that genuinely authoritarian movements will try to bridge that gulf.

But there is a much more insidious threat to our freedom even than that posed by a self-replicating and self-satisfied political class that lives in a kind of mentally-gated (and often physically separate) community. It is that posed by an important section of the intellectual class that values power much more highly than it values liberty: the liberty of others, I mean, for even Genghis Khan valued his own liberty.

It is difficult now to imagine a modern university intellectual saying something as simple and unequivocal as 'I disagree with what you say, but I defend to the death your right to say it.' He would be more likely to think, if not actually to say out loud or in public, 'I disagree with what you say, and therefore rationalise to the death my right to suppress it.' In public, he would be more circumspect, presenting a suppression of freedom as an actual increase in freedom, that is to say of real freedom, not the kind that leaves everyone free to sleep under a bridge. But he would know perfectly well in his heart that what he was after was power: the greatest power of all, that to shape, mould and colour indelibly the thought of others, a power to which he believes that he has a right by virtue of his superior intellect, training and zeal for the public good.

Of course, it might be argued that, in these days of Twitter and Facebook, what university intellectuals say and think is not so important that very much depends upon it any more, but I am not convinced that the days of cultural elites are over just because everyone is telling everyone else what he had for breakfast and his reaction to the latest news. The mediums employed by cultural elites may change, but not the fact of their existence; and it is difficult to imagine where else most of them will derive their ideas, of not from their university education.

Recently, I was reading for review a book by a woman, a 'resident scholar in the Women's Studies Research Center at Brandeis University,' about the problem of 'ageism' in America. It is certainly not difficult to find, in America and elsewhere, examples of reprehensibly bad treatment of old people, nor is it difficult to believe that such bad treatment (or neglect) is sometimes systematic. But what is so striking to me about the author's proposals for dealing with the problem is that she does not recognise that they conflict with freedom, and pose problems for the rule of law. To outlaw discrimination, for example, is to outlaw freedom, even if it can be decided without arbitrary assumption what discrimination actually consists of in any given situation. If I wish to employ someone but cannot hire whomever I choose, for whatever reason that I choose, whether good or bad, I am not free: I must hire according to criteria that are not my own. The author might certainly argue that her goals are ethically more important than that of freedom, that in fact fairness in one sense or another, in one field or another, is now more precious than freedom; but it is at the very least necessary to recognise that one is subordinating freedom to some other desideratum, or one will end with tyranny by default, as each enthusiast or monomaniac seeks to curtail freedom in pursuit of his favoured goal.

Very rarely do we find someone who is a university intellectual saying that 'x is indeed a desirable goal, even a highly desirable goal, but the cost to freedom of achieving it is simply too great.' It would be an excellent thing in the abstract if no one ever drank to excess (much less violence, cirrhosis etc.), but a system of surveillance in homes to ensure that no one did so would be odiously tyrannous. The author of the book to which I have referred would like to have all 'ageist' language expunged from films, radio, books, daily speech and even minds, on the grounds that many people have felt humiliated by it, that it reinforces stereotypes, and that stereotypes lead to bad treatment of the old. Even if this were empirically true (which might be doubted), what is being demanded as a principle here is language so anodyne that it could offend no one, lead to no stereotyping etc., for there is no reason to limit the cleansing of language to ageism. The attempt to rid the world of stereotyping is as totalitarian as it is in theory incoherent: for of course it relies upon the stereotyping of stereotypers, namely all of us. Show me a man without stereotypes, and I will show you a man in a coma. But mere impossibility has never stopped intellectuals from proposing their schemes.

I take the example of this book not because it is particularly bad, but because it is typical, one might say stereotypical, of a certain mode of thought that has become widespread in our societies. It is a threat to freedom, not that of mad dictators in fancy dress, but of termites.

27
Sewer Thing

T here has been so much written about the death of Osama Bin Laden that I do not think that I can usefully add anything to it. But I was very much taken by an article the day afterwards that appeared in the *Guardian*, the preferred newspaper of the British intelligentsia, by a columnist called Aditya Chakrabortty.

The drift and tone of the article may be gauged by the following:

Rather than winning hearts and minds, Berman [an expert on terrorism] suggests western powers should be focused on promoting sewage and schools.

And:

Islamic terrorism is deadly and unjustifiable. But it is also, at its root, prosaic.

The author's column is called *Brain Food*.

Now it is true that it is all too easily forgotten that, until quite recently, the most prolific suicide bombers in the world were not Islamic at all, but rather the militantly secular and Marxist (or marxisant) Tamil Tigers of Ceylon. This vile and deeply psychopathic movement received widespread support from a part of the Tamil population in London, who arranged many public demonstrations in its favour. The Tamil Tiger movement demonstrates that any analysis of suicide bombing that considers Islam as a necessary cause of it is mistaken.

But to suggest that young men and women blow themselves up, and hope to take with them as many people previously unknown to them as possible, as a kind of demand for municipal action is... well, beyond satire.

It is, of course, perfectly true that hygienic sewage disposal is highly desirable; that life without it is a good deal less aesthetically pleasing and more physically hazardous than it is with it, especially in crowded conditions, as the conditions of most people are. I do not think the point needs much elaboration. But Osama Bin Laden, as a qualified civil engineer and son of a billionaire building contractor, was surely in a unique position to improve sewage disposal in a large part of the world more directly than by resort to terrorism.

I have not done the calculation, but I suspect that there is not much correlation between the state of the sewage system and the number of terrorists that any society raises up. Young British Muslims who volunteer for terrorist activity often do not come from the prettiest towns of Britain, it is true, but even the worst towns in the country do not lack a functioning sewage system. It stretches credibility to the point of utter credulity to believe that young men and women are willing to die for other men and women's sewage disposal.

Whatever the real explanation for terrorist activity and behaviour (and no explanation of any phenomenon is final) it is certainly not this. Nor is it a lack of schools: terrorists, on the whole, are not completely unschooled, nor do they emerge from the most unschooled societies. I do not think that one kills oneself in such a murderous way in order that schools may be built and the little ones learn their ABCs. In fact, one has only to enunciate the idea to appreciate its absurdity.

Clearly the writer of the column is an educated and intelligent man, and so we are entitle to ask why he does not see at once that what he has written is laughable. Perhaps part of the explanation is the pressure of time; a columnist has to work to a deadline. Another explanation is that what he wrote goes against the opinion of most people: and to the intellectual (myself included), the unorthodox always has its attractions, irrespective of its truth or falsity.

But I suspect that the matter goes deeper than this.

There is a reluctance among intellectuals to believe in the depths of what I must call the human soul. I am by no means an unequivocal admirer of Freud, but at least he did not believe that the whole of life is contained on its surface, that Anna O., for example, behaved the way that she did because she wanted new curtains that no one would provide for her, or a larger monetary allowance. The human mind is a complex instrument and people do not blow themselves up for reasons as simple as that of a laboratory rat pressing a lever in an experimental psychologist's cage to obtain a pellet of food. (I always remember a wonderful cartoon in which such a rat says, 'Boy, have I got this psychologist conditioned. Every time I press the lever, he gives me a pellet of food.')

For people who think that other people blow themselves up for sewers (not sewage, as in the article, but one knows what the author meant), bad or even evil human conduct can and will be eliminated when some of the more obvious defects of the physical environment are ameliorated. Final victory over the human propensity to evil is in sight, therefore, and the road to heaven is paved with sewage-conduits – and/ or blackboards.

Dostoyevsky, thou shouldst be living at this hour, the *Guardian* hath need of thee.

I have notice a similar or cognate, but not exactly the same, tendency of intellectuals to refuse to face the possibility that ordinary people, those in no special position of authority, are capable of bad behaviour. They tend instead to explain it, or rather to explain it away, as if they did not want to recognise that it is often a response to the perennial temptations before mankind. They also fear that to admit that ordinary people can be morally defective is to be anti-democratic, or even an enemy of the people.

Not long ago, for example, I discussed the economic and financial crisis, with particular reference to Great Britain, with a young literary intellectual of intelligence. He was eager to accept the proposition that bankers had behaved with the greatest irresponsibility, making large profits from frenzied but obviously unsustainable activities. Bankers were for him precisely the kind of people with full moral agency, who could, and ought to be, criticised.

With slightly less alacrity, he accepted that the government bore responsibility also for what had happened. He found this more difficult, or less agreeable, to admit because it suggested that a change of government was necessary, but he hated the alternative. However, he did not feel able to dispute the fact that the government had benefited as much from the coining of fools' gold as had the bankers.

It was when I came to the part played by the ordinary population, neither bankers nor governors, that he bridled. But much of the population had behaved with gross improvidence and imprudence during the years of the bubble, and indebted themselves through a childishly incontinent desire to consume what they had not earned, at least not yet, and to speculate. And they did so, many of them, because they took a greedy delight in the inflating value of their houses, which made them think that they were growing rich. Their slogan was not think and grow rich, but sit in your house and grow rich.

As soon as I suggested that ordinary people were in part responsible for the crisis (though one could debate the share of the overall responsibility) my interlocutor grew excited. 'So,' he said, quoting the famous line of Brecht's, 'if the government doesn't like the people, change the people.'

I had said nothing of the kind, of course; I had merely said that many people (enough to make an economic difference) had displayed pretty ordinary human failings, of the kind familiar to most of us because we have at some time or other displayed them ourselves.

To take my own case alone: I have from time to time participated in feverish speculation, albeit to a limited extent, and have experienced the desire for wealth without the commensurate effort to obtain it. Sometimes I have paid the price, though sometimes I have also reaped some undeserved benefit. And I too have felt pleasure at the contemplation of the vertiginous rise in the value of my assets, though I was also aware that this was not a healthy sign as far as the larger economy was concerned, and might very well be followed by a fall almost as vertiginous. I was held back from boasting in public about the rise in value of my assets not by lack of greedy pleasure in it, but by the fear of appearing vulgar. But I would not be telling the truth if I did not admit that there was a song in my heart.

The crisis, moreover, has certainly strengthened my contentment with slow accretion, and a just return for prudence, providence and a willingness to postpone gratification – though, as I grow older, the question of how long such gratification should be postponed is beginning to rear its ugly head.

But to return to the question of intellectuals and their appreciation of human motivation. On the one hand they seem to want to deny the deeper currents that underlie the most extraordinary behaviour such as suicide bombing; on the other, they want to deny the quite ordinary or commonplace motivation for genuinely prosaic behaviour, such as spending too much. This is odd.

Perhaps they want to preserve the notion that man is by nature fundamentally good. In the *Guardian* article with which I started we are introduced to the concept of the altruistic suicide bomber, a concept derived from the fact that those who survived their attempted bombings claimed to be in pursuit of some altruistic end, when asked afterwards. It is surely very naïve to take this at face value: anyone caught in the middle of committing, and prevented from continuing, genocide would presumably argue the same thing. What the intellectual does not want to admit is that there is a joy in rage, and a joy in expressing that rage by evil deeds. And this is because we are constituted as we are: original sin, if that is how you want to put it.

Similarly, lust for easy gain is common enough. It is a human universal, not in the sense that it has always existed in all places and everywhere, but in the sense that it exists as soon as there is the opportunity for it to manifest itself. That is to say, there will be no final victory over such greed, and I have very little doubt that at some time in the not distant future there will be yet another epidemic of it, if we are not indeed in the midst of one already.

I do not mean by this the foolish view that we are all equally guilty of everything; and I hope I shall not be accused of spiritual pride or moral complacency when I say that at the height of the last speculative madness I did not give in to the temptation to borrow wildly in order to speculate, nor indeed have I ever done so to the extent of risking my all. (Nor, of course, will I ever achieve a fortune by means of speculation.)

The avoidance of the obvious is an occupational hazard for intellectuals, because the obvious threatens them with redundancy. One might have thought that it was perfectly obvious that there were deep psychological currents in suicide bombing, and equally obvious that there is widespread greed and incontinence during epidemics of speculative behaviour. Therefore it is only natural that intellectuals should be found who would argue precisely the opposite, that deep motives are in fact shallow and shallow ones deep.

Psychological realism is as important in ordinary life as it is in political life. Not to possess it may lead to serious consequences. In my own country, for example, the motives for criminality have been so mystified for so long by proselytising academics that efforts at repression have been, if not abandoned entirely, so weakened as to have turned one of the best ordered western societies into one of the worst within the space of a few decades, while at the same time reducing many of its civic freedoms. The motives of criminologists are far harder (on the whole) to discern and understand than those of thieves.

28

The Baseness of Acid

R evenge, said Lord Bacon, who was not himself completely foreign to the impulse, is a kind of wild justice, which the more
man's nature runs to, the more ought law to weed it out. Furthermore,
he says, it does the revenger harm, psychologically, for he goes on to say:
This is certain, that a man that studieth revenge keeps his own wounds
green, which otherwise would heal and do well.

It is my observation from many cases that he who seeks revenge, or
merely compensation, through our system of tort law, 'keeps his wounds
green,' sometimes for many years or even forever. Not only does it take
years to obtain that revenge in the form of financial redress, but it is
seldom of a quantity sufficient to satisfy the revenger, who then feels
doubly wounded, first by the wrongdoer and second by the legal system;
or if the redress is sufficient, the revenged man has to keep his wounds
green in order to justify to himself the large sum that he has procured.
He does not want to admit to himself that money is a healer, for that
cheapens his suffering and suggests that he was mercenary in the first
place.

But the *lex talionis* – an eye for an eye and a tooth for a tooth – appeals to man's instinctive feeling for justice. Note that Lord Bacon does
not say that revenge is unjust: he says that it is wild, in other words that
it has a tendency to get out of hand and become a source of new injustice. Interestingly, he does not say in his little essay on the subject that

revenge, in its best aspect, acts as a deterrent to future acts of injustice; he says rather that, where it is least harmful, 'the delight seemeth not so much in doing the hurt as in making the party repent.'

This, it seems to me, is rather over-optimistic about human nature and its sometime joy in inflicting pain on others; but certainly Bacon's essay is an early example of the shift from shame as the regulator of behaviour to guilt as its regulator.

A very interesting recent case of the operation of the *lex talionis* comes from Iran. A woman called Ameneh Bahrami had acid thrown in her face by a spurned lover, a man who wanted desperately to marry her but was several times refused, threatened to kill her and eventually decided to ensure that, if she would not marry him, she would marry no one else. (I often heard these words from a murderer: If I can't have her, no one else will.)

The acid that the spurned lover threw in her face disfigured her terribly and blinded her for life. In the Middle East and South Asia this behaviour is common: there are thousands of cases a year, or so we are told.

The victim insisted upon her legal right under Iranian law to retribution; and the retribution she demanded was that the perpetrator be blinded in the same way. The court eventually agreed, with a significant alteration: the man was to be blinded by acid under anaesthetic. There has been an international outcry of the usual kind; Amnesty International called the punishment a form of torture, and western governments have protested. For once, the Iranians have – at least for the time being – taken notice of the outcry, and postponed the execution of the decreed punishment.

What is wrong with it? It cannot be said to be unjust in itself, disproportionate to the original offence, which was one of the greatest magnitude. Can one honestly say that the perpetrator did not deserve such a punishment (assuming that he was not insane when he committed the act, a subject of great complexity worth going into in itself). His crime was not only premeditated and highly organised, it was one consciously designed to bring about a maximum of suffering – decades of suffering, to the very last day of her life – to the victim. It is difficult to think of the crime without rage, or to imagine the victim without sorrow.

She does not deny her thirst for revenge, and which of us is in a position to condemn her for that? But she also says that, in a society in which many women are horribly treated by men who get away with it, this punishment will be both exemplary and deterrent. She does not want others to suffer as she has suffered; and again, no one could blame her for that. Whether the punishment would actually be a deterrent remains to be seen (in any case such questions tend to be undecidable), but it is certainly not implausible.

For all this, however, I think most of us would bridle at the punishment. Our objection to it would be not be that it was unjust, but that it was brutal. Justice is not the only virtue in life, and justice must therefore be tempered by mercy.

Here let me say that mercy is not the same as forgiveness. The law can be merciful but not forgiving: often it cannot inflict what a man deserves because what he deserves is too awful to contemplate. Similarly, I may forgive a wrongdoer for what he does to me, but I have no right to forgive him for what he does to you: only you have that right, a right that, incidentally, is not a duty. For me to forgive the person who wrongs you is a form of self-indulgence or, worse still, moral exhibitionism. Hence, no one, and certainly not the law (whose duties are different from those of individuals), has the right to forgive the perpetrator of the crime in Iran except the victim, and she chooses, very understandably and in my view rightly, not to do so.

The punishment is nevertheless brutal because the deliberate infliction of such an injury is brutal irrespective of the motives for inflicting it. Can one imagine oneself being prepared to drop acid into the eyes of a sighted man in order to blind him, however evil one might believe him to be? Would one trust oneself to be doing it for good rather than for bad reasons? Not every punishment is permissible that is just and proportionate.

What of the utilitarian argument, that the end would justify the means? Let us suppose that blinding the perpetrator would deter a hundred similar acid-throwers (assuming such a thing could be known indubitably in advance, which itself is highly doubtful). Would that make the punishment right?

Let us suppose also that the hundred acid-throwers deterred would disfigure but not blind. Is there any way of reducing the suffering caused by the blindness of the punished perpetrator to the same units as the suffering caused by the disfiguring avoided, such that a genuine profit and loss calculation can be made? I doubt it.

Further, let us suppose that the perpetrator to be blinded is not the real perpetrator, but that everyone supposes that he is, such that the deterrent effect of blinding him would be the same as if he were the real perpetrator. Would that be all right?

Surely the answer is No, it would not be all right, though the net effect of the punishment might be beneficial. It would not be all right because it would be grossly unjust. Nor would matters be rectified, morally, if the gross injustice were never to be revealed. Utilitarianism as a general account of morality must therefore be unsatisfactory, even if it is often useful and we all sometimes use it in our moral thinking.

Do circumstances alter cases? Could the blinding of the perpetrator be right in Iran but wrong in the West? After all, acid-throwing in the faces of women seems to be not much of a problem in the West, but it is a problem in the Middle East and South Asia. It does occasionally happen here, but not on a scale to be what one might call a sociological phenomenon. A punishment that appears brutal and even disproportionate in one set of circumstances might therefore not appear brutal or disproportionate in another.

Here I confess to facing both ways, or to not being quite able to resolve the dilemma. Where people behave well in a certain respect, it is not necessary to threaten them with condign punishment in the event that they fail to behave well, at least not from the point of view of deterrence. (Or is it the threat of condign punishment that ensures that they behave well in the first place?) But where they do not behave well, where they behave badly, the threat of condign punishment might produce an improvement.

A few years ago, while working in a prison in England, I met a prisoner on remand who was accused of throwing acid in the face of his girlfriend who wanted to leave him. The allegation was that he did so to disfigure her so that no one else could 'have' her.

At first he denied the accusation, on the general grounds that 'I don't do them kind of things.' Later I asked him whether he had ever been in prison before, whereupon he said that he had. It turned out that this had been for throwing ammonia in the face of a girlfriend about to leave him. Perhaps he regarded throwing ammonia in the face of a woman as a completely different kind of thing – as in 'them kind of things' – from throwing acid in the face of a woman, a difference that justified his initial denial; but in the end, he confessed that he had done it, though only while drunk.

He received a prison sentence of two years, of which he would serve but one; he would be out, ready to throw more acid or ammonia in the face of a woman, in only twelve months. I felt this to be a completely unjust and indeed outrageously inadequate sentence, although as far as I know he disfigured the women 'only' rather than blinded them.

However, it takes little effort of the imagination to understand the terror of these women, to say nothing of the misery of their disfigurement; and if, in distinction to the Iranian acid-thrower, he failed to blind them, it was not through any scruple on his part. A man who throws acid in the face of a woman cannot be said to care very much whether or not he blinds her.

Honesty compels me to admit, however, that the extreme leniency with which (as it seemed to me) this man was dealt by the courts did not lead to a local outbreak of acid-throwing in the faces of women. There was plenty of violence against them, but not of the specifically acid-throwing kind.

All the same, I could not help but feel that a grave injustice had been done: for unwarranted leniency is also injustice. If I had been one of the women in whose face he had thrown acid, or even the relative of one of the women, I would not have thought that the state had taken seriously either the crime itself or the suffering caused by it, or indeed its duty to repress such crimes by signalling that they would not be tolerated. Apart from anything else, it did not seem to me that throwing acid was so *sui generis* a crime that it should be considered different from other forms of violence and therefore treated differently from a penological point of view. Tolerance of acid-throwing was, in effect, tolerance of all kinds of violence.

The leniency of the sentence given to this man called forth no expression of public outrage. Amnesty International did not protest, nor did other governments, arguing that the English law was failing to protect people as it should. We have so far been affected by sentimentality and a lack of realism that only undue severity moves or appals us, rarely the opposite (and then only for reasons that require special pleading).

But a foolish consistency is the hobgoblin of little minds, of. With consistency a great soul has nothing to do; and if we are inconsistent, it is proof enough that we are great souls.

29

The Rape of Innocence

My sister-in-law phoned to tell me that Dominique Strauss-Kahn had been arrested: the telephone being my only means of communication with the world from my tiny corner of *la France profonde*. She telephoned me again yesterday to tell me that the case against him was on the verge of collapse. If it does collapse, a lot of people will have to re-arrange their memories to demonstrate, both to themselves and to others, that they had been right about the whole business from the very beginning, and had always smelt a rat.

For myself, I think I can safely say that, though I did not really know exactly what to think, I did not believe in the story that Strauss-Kahn emerged from the shower all ready to go, and jumped upon the chambermaid without a preliminary word of seduction: at least not unless he had a tumour in, or dementia of, his frontal lobes, and was therefore not responsible for his actions.

Most people found it difficult to suspend their judgment: they either thought him fully innocent or fully guilty, according to their predilections. Oddly enough in France there were those who thought him guilty of the act and yet defended him, arguing that his preposterous behaviour (for such it would have been if it had been as it was alleged to have been) a kind of human sacrifice or even martyrdom, that is to say a way of avoiding his own fate, which appeared at the time to be that of becoming the next President of the French Republic. He did not really want power, and so organised his own downfall. It is amazing what thoughts the necessity to fill column inches will stimulate in journalists and commentators.

But most of those who assumed his guilt did not in the least admire what he was alleged to have done, far from it: and, indeed, it is difficult, without the kind of extravagant mental contortions habitual among certain French intellectuals, to see anything other than the reprehensible in his alleged, but unproven, behaviour. Various feminists on both sides of the Atlantic started to demonstrate against the type of behaviour of which, they said, Strauss-Kahn's alleged behaviour was emblematic. They generalised wildly from events that had not yet been proved to have taken place.

It is extremely unlikely that they will ever apologise for their assumption of his guilt (here, of course, I am making my own assumption that the prosecutors' discovery that Strauss-Kahn's accuser is the very opposite of a reliable and credible witness is itself correct, and also that Strauss-Kahn will be exonerated of the worst charges against him); for, they will say, though Strauss-Kahn be as innocent as a sucking dove, yet what they were protesting against, or drawing our attention to, was a real phenomenon, namely that of sexual harassment and the abuse of power by men in positions of authority. The fate of Strauss-Kahn himself does not interest them; the chambermaids who demonstrated in New York were, presumably, protesting about what they had themselves suffered, hitherto in silence. (It is, after all, nearly a century and a quarter since the French writer, Octave Mirbeau, published his novel, *Le journal d'une femme de chambre*, full of Strauss-Kahnian stories.)

Now it is very unlikely that, in response to the allegations that the prosecutors themselves have now made against Strauss-Kahn's accuser – among them, falsifying her application for asylum in the United States, money-laundering and what amounts to attempted extortion – will result in any public protests against the presence in the United States (as well as in Europe) of large numbers of bogus and criminally-connected asylum-seekers from many Third World countries. Strauss-Kahn's accuser was described as a quiet-living, law-abiding, domesticated, humble and modest person, and now stands revealed – again if the accusations of the prosecutors are true – as an associate of criminals and drug-smugglers.

Now Strauss-Kahn stands in exactly the same relationship to powerful men as a whole as does his accuser to asylum-seekers from countries such as Guinea as a whole. Yet it is deemed morally and intellectually respectable to demonstrate in the one case but not in the other. Why?

It might be said that powerful men as a group are not worthy of much sympathy: they can look after themselves, hire expensive lawyers, afford to pay $250,000 a month to keep themselves out of prison etc. But this is beside the point: a powerful man is not guilty of anything just because he is a powerful man (unless all power is regarded as criminal, an absurdly adolescent and utopian point of view). He has the same right to the presumption of innocence as anyone else.

In other words, if it was reasonable and justified to demonstrate against the ill-treatment of chambermaids because of what Strauss-Kahn allegedly had done, it would be reasonable and justified to demonstrate against the presence of bogus asylum-seekers because of what his accuser allegedly had done.

Again, it might be argued that while one social problem is serious – that of the abuse of subordinate women by powerful men – the other, that caused by bogus asylum-seekers, is not. But this, surely, is very far from certain.

We don't have any reliable statistics about the scale of either problem; self-reports of victimisation, for example, are suspect for obvious reasons. And anyone who has lived in an urban area where bogus asylum seekers gather is unlikely to underestimate the social costs of their presence, for example in crime and trafficking of all sorts. This is so even if - as I believe - many bogus asylum-seekers live otherwise decent and law-abiding lives. Incidentally, the idea that the traffickers would turn into decent, respectable, hard-working citizens if only the drug trade were legalised strikes me as naïve, to put it no higher.

There is no common measure allowing us to say which of the two is the worse problem, but it seems unlikely that we can say for sure that, *grosso modo*, one is serious and the other trivial. In other words, there is no purely objective or intellectual reason why it should be morally acceptable to demonstrate in the wake of the arrest of Strauss-Kahn, but not in the wake of revelations about his accuser. It is more a question of sensibility than of rationality.

Here I do not exclude myself. I would find it distasteful if there were now demonstrations against Guinean or West African bogus-asylum seekers on the grounds that some of them laundered money, dealt in drugs, made false allegations, took advantage of legal provisions to enrich themselves illicitly, etc. The reason for this is that – apart from personal reasons, being myself the son of an asylum seeker, and having had many asylum-seekers as patients – anti-immigrant feeling is often associated with that most primitive, durable and strongest of all political emotions, hatred. Xenophobia lies not very far below the surface in most of us, can easily be aroused and needs to be guarded against.

I did not feel quite so repelled by the feminist demonstrators. I thought they were wrong, unfair, and possibly prejudicial to the chances of a fair trial of Strauss-Kahn; they were unattractively shrill; but perhaps my reaction against them should have been stronger.

A pleasant pastime for intellectuals is the game of estimating whether the religious or the anti-religious have done more harm, for example have killed more people or censored more books. Needless to say the anti-religious favour one answer, and the religious the other. It is not an easy question to answer definitively, because the religious have been at the game for so much longer than the anti-religious, though the 'productivity' of the latter, as measured by corpses per year of power or influence, for example, is far greater. Perhaps the answer is a draw.

Now when it comes to the hatred of powerful men, I think it fair to say that it has done a great deal of harm likewise. Few are the revolutionary Saturns who have not devoured the children not only of the powerful, but their own children, and the children of many others beside. This is a proposition that hardly needs historical proof; and it is the case even where the powerful men against whom the revolutionaries revolted were themselves no angels, as was usually – indeed, always - the case.

Hatred of the powerful, supposedly because of love of justice, is therefore not in itself a noble or a good emotion. In essence, it appeals to the same baseness, and calls for the same low, primitive and visceral reactions, as hatred of foreigners and immigrants. Another interesting parlour game for intellectuals might be to estimate whether xenophobia has been responsible for more or fewer deaths in the twentieth century than the hatred of the powerful, taking xenophobia and hatred of the powerful in a broad sense, to include the Jews on the one hand and the rich on the other. The result would be, as the Duke of Wellington said about the Battle of Waterloo, a damned close-run thing.

The gutter press, in displaying the privileged, immensely rich and clever Strauss-Kahn taking the *perp walk*, appealed to the lowest instincts of the mob. It was the modern equivalent of the public execution, that entertainment that used to draw such crowds in Europe (and would still do so, if permitted). There was quite a lot of self-congratulation that Strauss-Kahn was not treated better or differently from all the other presumed criminals in New York, thus demonstrating that the law in

America is the same for all, and that the depiction of justice blindfolded is justified.

This rather overlooks the fact that an equal injustice is not the same as equal justice, any more than cruelty towards all is a form of kindness. Even to call it the *perp walk* is an injustice, since those who take it are, in the eyes of the law, still innocent men. It might well be true – let us hope, at any rate, that it is true – that the great majority of those arrested by the police and accused by the prosecutor are in fact perpetrators, and guilty of the crimes they are charged with; but still no one is to be presumed guilty until proven innocent, and therefore the *perp walk* is an offence against justice.

I am told by a New York friend that it was instituted as a punishment in itself for people whom the police and prosecutors knew to be guilty, but whom they also knew that juries would acquit on grounds of racial solidarity (that solidarity apparently never extending to the victims of their crimes, who are frequently of the same race). But, as we used to say when we were young, two wrongs don't make a right. You do not correct an informal injustice by committing an official one.

Likewise the judge appeared to make a gross mistake when Strauss-Kahn appeared in court the first time and was remanded into custody. One of her reasons for not granting the accused bail was that he had been apprehended while waiting for his plane to take off to Paris. She said that he was fleeing, a word which contains a very strong presumption of guilt, a presumption which it was her most elementary duty not to make. The facts actually supported the presumption of innocence, for Strauss-Kahn, who had booked his ticket well in advance, and could therefore be construed as merely keeping to his schedule, had to be tricked off the plane because he enjoyed diplomatic immunity while on it. The fact that he agreed to get off suggests (though, of course, it does not prove) that he did not think that he had anything to fear, and that he thought himself an innocent man.

Actually, the decision to release Strauss-Kahn from his stringent bail conditions while the charges against him were still maintained was completely illogical, and meant either that the conditions had been wrongly imposed in the first place or wrongly withdrawn, for his capacity to abscond remained unchanged in the meantime. The only thing that had changed was the prosecution's confidence in its own case: and it is now the DA of New York's turn to fight for his name and career. But to say that the prosecution's declining faith in its own case made such a difference to the risk of Strauss-Kahn absconding that it was now safe to free him was itself a slur on his character and that the presumption of his innocence had been a fiction.

I think the only thing that it is perfectly safe to say about this case as a whole is that there are, even as I write this, a hundred scribblers scribbling their books about it. Indeed, I recently bought the first of the crop in France which, however, read as though it had been written by a committee rather than by one person. Its most valuable chapter was a non-exhaustive collection of the foolish things people had said about it. A French comedian called Olivier de Benoit joked:

> DSK is simply the victim of the speed of the American system of justice. He had hardly put his hands on the chambermaid than he was behind bars. In France, it would have taken at least two years for the justice system to have charged the chambermaid with disobeying a high official.

30
Who Is to Blame?

S ome years ago I had a patient who kept all his appointments dressed in military costume, including boots. His camouflage made him highly conspicuous in the streets of the city, but he was obviously a soldier in the way that Marie Antoinette was a shepherdess. He no more wanted to live under military discipline than Marie Antoinette wanted to herd sheep. On the arm of his shirt was sewn a little West German flag, and I could not help but wonder whether this was a metonym for something rather more sinister.

He was a great lover of animals, as were many leaders of the Nazi Party and as, indeed, am I. He was so incensed against meat-eaters, he said, and about the cruel conditions in which meat is produced, that he often felt like shooting the people at the meat counter of his local supermarket. To this end he had joined a gun club, for it takes practice to kill carnivores selectively in a crowded supermarket.

I did not think his threat an entirely idle one; he was one of those people for whom the love for anything is always accompanied by a countervailing hatred of something else. He caused me considerable anxiety, not the least part of which (I am not proud to say) was the public opprobrium to which I should be subjected were he to open fire and kill people *en masse*: for everyone would ask why I, supposedly in charge, had done nothing to prevent it? In a vain attempt to spread or share the public responsibility for what he might do, I took legal advice: was it my duty to inform the police or to maintain patient confidentiality (I hoped the former)? The advice might be different today, after so many terrorist incidents and mass killings, but I was told then that I had no grounds for going to the police. The burden of anxiety was mine alone to bear.

I am glad to say that he did not carry out his threat; but would the moral problem of the treatment of animals raised for meat have ceased to be a real one had he done so? If he had shot dead twenty people in a supermarket, say, ostensibly to improve the treatment of animals raised for meat, would the monstrosity of his action have meant that we would now be entitled to disregard the question altogether?

It is surely a moral duty (if anything is a moral duty) not to inflict avoidable suffering on sentient beings, if that suffering is inflicted merely to give us some later gratification. If we caught someone inflicting pain on a cow or a pig, or even a chicken, merely for the sadistic pleasure of doing so, we should be appalled and rightly so; therefore we should not shut our eyes to the conditions under which meat is raised so that it should be easily and cheaply available to us all. To this extent, the animal rights activists are justified.

Of course, the problem is not an absolutely straightforward one. Most farm animals owe their very existence to the fact that they are of economic value to the farmer; without that value, much of it derived from our carnivorous habits, they would have had no existence and therefore no progeny. At what point it is better for a creature not to have existed at all than for it to have the life that we think is horrible, is not easy to determine. Most people do not commit suicide in the very worst of circumstances; this may be because they believe that life has an intrinsic value that outweighs the experiential content of that life, or because they are afraid of death, or because they retain some hope for the future. None of these feelings or beliefs can be attributed to the consciousness of animals, but the fact is that most creatures flee or resist death as if life still had some value for them. We cannot be absolutely sure, therefore, that it is better for a certain pig never to have lived at all than to have lived the life that it actually did live. We can, however, be pretty sure that it suffered.

For most of us it is self-evident that the mistreatment of animals could not possibly justify killing people, whether or not at random, and whether or not they were personally responsible, directly or indirectly, for the mistreatment. One of the first moral precepts that any child learns is that two wrongs don't make a right, *a fortiori* if the second wrong is greater than the first. And if a greater second wrong is committed, it does not mean that the first wrong has thereby ceased to be a wrong.

It follows, therefore, that even if my patient had acted out his fantasy, even if he had committed the monstrous crime that was in his mind to commit, the moral problem of the treatment of animals would not have been solved. And the fact that there are monomaniacs who look at the whole world through the distorting lens of that problem also does not mean that it ceases to be a problem.

Even monomania has its uses or benefits as well as drawbacks, not only for individuals, but for society as a whole. Monomania answers the difficult question for individuals of what life is for: life is for the pursuit of whatever goal the monomania suggests is desirable. Where this goal is harmless or beneficial, the monomania is harmless or beneficial; no doubt it is an excellent thing for society that a variety of learned or inventive monomaniacs till their tiny fields to the exclusion of all others, for otherwise those fields might not be tilled at all, at any rate not so well-tilled. As for myself, I think my wife would say that I am a serial monomaniac, obsessed – though for a relatively short time on each occasion, from one month to three months – by a single subject upon which I happen to be writing something. The danger of monomania is when a single idea not only crowds out other subjects from the mind, but appears to the monomaniac to be of unique and even sole moral importance.

It is also important to remember that the psychological provenance of an idea, or the political circumstances in which it is propounded, does not determine its truth or validity. The dangers of smoking, for example, were first appreciated in Nazi Germany, where the first epidemiological research into those dangers was done. (This is all recorded in Robert Proctor's excellent book, *The Nazi War on Cancer.*) Indeed, one of the most eminent post-war British researchers into the connection between smoking and lung cancer, among many other diseases, Richard Doll, went to Nazi Germany in his youth and listened to lectures there on the subject. He was subsequently somewhat coy about the origin of his ideas, either though amnesia or fear that the origin might contaminate them in the eyes of his colleagues (or some combination of both, the human mind being a marvellously subtle instrument). But the fact is that smoking *is* harmful to the health; and the fact that it was doctors working in Nazi Germany who accepted, *ex officio* as it were, the removal of Jewish doctors from the profession, does not alter the conclusion.

All this is but a prolegomenon, not to a confession, exactly, but to an acknowledgement of unease: for a close acquaintance of mine informed me that I had been quoted, albeit indirectly and quite possibly inaccurately, by Anders Breivik, in his 1500 page manifesto, posted on the internet shortly before he went on his terrible rampage in Norway. I was quoted only *en passant*, and much less than many others (some of my acquaintance); but no one likes to be thought of as even a remote inspirer of such a man or of such an action.

Needless to say, I am clear in my mind that nothing I have ever written could be taken as a justification for, much less an incitement of, mass killing. No minimally sensible person could derive a reason for such a killing from my words. But this does not entirely dispel my unease, precisely because I have spent much of my writing career propounding the view that ideas have consequences, and that many contemporary undesirable social (or antisocial) phenomena are caused not by 'objective' economic or physical conditions, but by the ideas that people have in their minds, often instilled into them by intellectuals without much thought of their adverse consequences if taken too literally or distorted.

Is it possible, then, that by emphasising the less attractive aspects of modern society and culture, by repeatedly drawing attention to the deleterious social and psychological effects of welfare dependence, by criticising multiculturalism as a doctrine and as corrupt bureaucratic opportunism, I may have contributed, if only a mite, to the poisonous, paranoid, narcissistic, grandiose and resentful brew in the mind of Breivik, who took what I wrote, even if at second-hand, in completely the wrong way and drew ludicrous but murderous conclusions from it? And if I did contribute that mite, does it mean that I should now retire into guilty silence, lest there be other Breiviks in the world?

In writing on the subject of immigration, for example, I have always felt an undertow of anxiety and guilt, not only because I am myself the descendent of a long line of refugees, but because I know that this is a subject on which the vilest passions and basest emotions are quickly aroused. There is, after all, a long history of such vile passions and base emotions in many, perhaps in most, countries. Thus to say anything about mass immigration other than it is an excellent thing is potentially to give intellectual succour to some very nasty people.

But while history provides us with analogies, they are never exact. As human beings, we are condemned – it is both our glory as well as our burden – to live in perpetual near-novelty, and therefore to have to make continual leaps in the dusk if not in the total dark. We cannot treat the present as if it were a mere repetition of the past. To be mesmerised by precedent is as foolish as to take no notice of it whatever. It is said that generals always fight the current war with the strategy and tactics of the last; in like fashion, social commentators and reformers are reluctant to let go of past problems in favour of the problems that confront them now. A phenomenon – immigration – can keep its name while changing its nature; and it is obvious that the social consequences of immigration depend on the qualities of the immigrants as well as on the quality of the society into which they immigrate.

To be reduced to silence on an important subject, to decree in effect that only one opinion on it may be openly expressed, for fear of filling the minds of the unstable with murderous resentment, is to place a great deal of subject matter *hors de combat*. It is true that as yet no climate activist has killed people, and that 'only' three bank employees lost their lives in the riots in Athens (and probably not by the direct intention of the rioters at that), but there is no reason to suppose that extreme climate activists or protesters against finance capitalism are, and must forever remain, immune from murderous impulses. The human mind is capable of finding a *casus belli* in almost anything, and of rationalising violence when it wants to commit it. If an environmental activist were to act in imitation of Anders Breivik, I should not blame those who warned against global warming, nor even Anders Breivik himself.

As all who have ever suffered from anxiety know, however, rational considerations rarely soothe it away altogether. And if I were really free of it in this instance, it would be a matter of indifference to me that Breivik had referred to me in his preposterous manifesto; but it isn't. I will have to live with that anxiety, as I once lived with the anxiety that my patient might mow down shoppers in the supermarket.

31
The Meaning of Pyongyang

There are some countries that, once visited, retain a dispro-portionate hold on your imagination. Among them, for me at least, are Haiti and Liberia, two small states that are known to the world at large principally for their political, and sometimes for their natural, catastrophes. They are marginal from the point of view of the world economy, I need hardly say, and yet their history has something about it that makes it seem significant beyond itself. No one, I think, can study the early history of either country without being moved by it; and just as the biography of a single person can also be a portrait of an age, so the history of an otherwise insignificant country can tell us something important about the human predicament as a whole, for example our tendency to turn liberation into a new form of servitude.

North Korea is another country that, once visited, is not easily for-gotten. Its hold on the imagination, however, has nothing of affection in it, as does that of Haiti or Liberia. This absence of affection is no reflection upon the Korean people, but rather upon the political system that reigns there. Spontaneous contact with Koreans is precisely what the regime attempts at all costs to prevent, and succeeds to an extent unique even for the communist, or formerly communist, world. Com-pared with North Korea, Hoxha's Albania was a free country. In short, North Korea has all the fascination of sheer horror.

More than twenty years after I visited it as a member of the British delegation to the International Festival of Youth and Students, though at the time I was neither a youth nor a student, I still buy books about North Korea whenever I see them, which is not very often. The International Festival was a four-yearly jamboree of communist youth held in a different communist capital, and this was the last ever to be held. It would take too long to explain how I was selected to go; but I was disappointed not to be among the elite of the British delegation to march past the Great Leader, Kim Il Sung, in our uniform which, curiously enough, consisted of a brown shirt. A sense of irony is not one of the qualities to be expected of true believers in the North Korean version of paradise.

I happened quite recently to see a review of a book about North Korea in the literary pages of *Libération*, the French newspaper that was once very left wing. It was of the French translation of B. R. Myers' *The Cleanest Race: How the North Koreans See Themselves – and Why It Matters*. I ordered it, and for some reason it took a little while to come.

I knew of Myers only for a splendid and enjoyable little polemic that he wrote over ten years ago about the unreadability of contemporary American literary fiction, called *A Reader's Manifesto*. I noticed at the time that he was by profession what one might call a Pyongyangologist, that is to say a student of North Korea, who taught (and still teaches) in South Korea. A strange combination, literary critic and Pyongyangologist, but why not? It often seems to me that the best literary critics, especially nowadays when the careerist temptation to write obscurely and uninterestingly is so strong, have other jobs as well.

Myers is clearly a formidably clever man. American by birth, he grew up partly in Germany, where he studied Korean. He is obviously polyglot, and not just in closely-related languages. He brings to his book what very few people beyond the borders of the hermetic state can claim to have, namely a knowledge of its film and literature. And, what is even rarer, he is one of those people who have studied North Korea without becoming an apologist for it (it often takes a strong ideological predilection to undertake studies as arduous as those he undertook). There is hardly a single word in his book that could be construed as being sympathetic to the regime.

He knows infinitely more than I about North Korea, and I know more than the average person walking down a western street; and yet, for all my admiration for him, I find myself not in complete agreement with his analysis.

If there is a central idea to his book, it is that the nature of the North Korean state owes very little to Marxism, and far more to Japanese colonialism of the period 1905 – 1945.

To be sure, I don't think anyone who read Marx and knew nothing of the development of Marxist states inaugurated by the Russian Revolution would deduce anything in his imagination resembling the North Korea of the egregious Kim family. But I do not think this settles the matter.

Because the North Koreans (like the South Koreans, apparently) have a strong racial consciousness, seeing their own supposed racial purity as an advantage over more mongrel populations, Myers suggests that fascist Japan was actually more important as a model for North Korea than any Marxist theory or practice. I think this is mistaken, for a number of reasons.

First, North Korea was instituted as a state by the Soviet Union. Even now, its military uses Soviet-style uniforms. The very cover of the book (in its French edition) uses a picture of Kim Il Sung in a uniform that Marshal Stalin could have worn in his salad days. The omnipresent iconography of North Korea is a mixture of Soviet and Chinese, but the pictures of the Great Leader standing in the midst of a field giving peasants advice about how to grow corn (or whatever) while an amanuensis takes his every word down in writing, is clearly of Stalinist origin.

The uniformity of proletarian dress that North Korea imposed was clearly more communist than Japanese colonialist or fascist of any other stripe; and its architecture is clearly Stalino-Maoist. In any case, it is mistake to suppose that fascism and communism were or are polar opposites in all important respects. The iconography of the Nazi variant of fascism and of Stalinism are sometimes difficult to distinguish, as Igor Golomstok demonstrated in his book on the subject.

Myers refers in his book to the collectivisation of agriculture in North Korea, a policy completely normal for a communist regime but not for a fascist one. This is not a small matter, for collectivisation was a major cause or precondition of the famine that killed at least a twentieth of the North Korean population, and also for the chronic food shortages in the country. Indeed, such collectivisation is of immense cultural and psychological significance, as well as economic. There is nothing fascist about it.

Under fascism, capital continued to be privately owned but on condition of its owner's total obedience and loyalty to the state; North Korean capital is not privately owned even in this sense. The difference is important, both economically and culturally; and, again, North Korea is clearly more communist than fascist.

It is perfectly true, of course, that the regime's imposition of ideological conformity, its ruthless suppression of all opinion other than its own, could just as well be fascist as communist. But in fact, there are good grounds in Marxism for such totalitarian control. I am not speaking here of Marx's personal intolerance of any opinion other than his own, remarked upon by his own father as a psychological quirk from a very young age, and by virtually everyone else who ever came in contact with him; he could not think of anyone who disagreed with him on anything as other than philistine, stupid, dishonest etc. No; I refer to his dictum that it is being that determines consciousness and not consciousness that determines being. If this is the case, you can change consciousness by a change of being; those who have different consciousness from yourself must have it because they have a different being (ie economic interests) from your own, and must therefore be enemies, since in Marxist sociology all difference is enmity; and infinite manipulation of being is justified in order to produce the New Man, that human who will finally be truly human because of his communist consciousness. It is, as the Marxists themselves would say, no accident that all Marxist regimes have ended up as totalitarian dictatorships.

Myers would appear to be on better ground as far as the racism and xenophobia of North Korea are concerned, which at first sight are more reminiscent of Japanese imperialism and fascism or Nazism than of Marxism. There is clearly an internationalist strain in Marxist doctrine that rejects all racism and xenophobia: the proletarian, after all, supposedly has no homeland, and a Mongolian proletarian is supposed to feel an immediate and spontaneous class solidarity with a Costa Rican one.

But even this needs some qualification. Marx himself was a racist, a ferocious (and lifelong) antisemite, though Jewish. He and Engels spoke of people as unhistorical and therefore unimportant, and backward for racial, that is to say biological, reasons. The historical mission of these nations was to disappear from the face of the earth. Genocide is not completely alien to this way of thinking.

Moreover, the structure of Marxist thought is highly propitious to hatred of whole collectivities of the human race. One might even call socialism the antisemitism of the intellectuals. Once you start hating whole classes of the human race, it is difficult to know where it will end. Certainly, xenophobic nationalism was not alien to communist regimes other than the North Korean: the Cuban, Albanian and Romanian regimes were all dab hands at it, for example. The sheer difficulty of loving your class peers in distant countries renders genuine Marxist internationalism psychologically unlikely, not to say impossible. Hatred is a much stronger political emotion than sympathy, let alone love. In sum, I do not find the racism and xenophobia of North Korea so very surprising, or evidence of the regime's Japanese fascist rather than communist inspiration.

There are other arguments in the book that I do not find convincing. The personality cult of Kim Il Sung is certainly not an anomaly if you consider communist regimes as a whole. The fact that the Great Leader had read little or no Marx (if, indeed, he had not) is not evidence of very much; one can become a Marxist by reading a primer of Marxist philosophy and economics. Marxism is a climate of opinion as well as a doctrine; millions of people now adhere to John Stuart Mill's doctrine of liberty without having read it, or even without having ever heard of him.

There are other things that troubled me in the book. At one point, Myers makes much of the true fact that no dictatorship can survive without a degree of co-operation and acceptance by its population: that no dictator can do without his henchmen, that no regime can survive without the support of many people. But when he tells us that it is not the normal terror of a police state that sustains the North Korean regime, he seems to me to be going too far. I remember an incident when I was in Pyongyang in a car with a fellow-delegate. We stopped for the motorcade of the Great Leader which was passing in another direction. My companion took out his camera to take a photo; our driver, until them as impassive as a stone wall, turned to him and let out a scream. The look of terror on his face was such as I shall never forget; I have never seen anything remotely like it. Behind that look of terror, of course, there must have been knowledge of what might happen to him (and perhaps to others) in case he broke the rules, one of which was evidently never to point a camera at the Great Leader. Only official, that is to say doctored, photos were allowed.

The very subtitle of the book (for which, of course, the author might not be responsible) is troubling, for in a situation such as the North Korean it is surely impossible to estimate how the people see themselves. Official productions (and all productions there are official) do not help; at best they are evidence of the point of view that the political system wishes to inculcate in the population. Whether they succeed in doing so is something we shall have to wait for the end of the regime to find out, and even then the honesty or dependability of the testimony might not be beyond reproach. There were not many convinced Nazis after the war, and perhaps there will not be many convinced believers in Kim-ism after the regime collapses.

The interpretation of a country or society is never final. When Myers says that North Korea owes more to its Japanese colonial past than to communism, incidentally denying a premise of Korean nationalist historiography that there was a complete and utter opposition between the Koreans and their Japanese occupiers, I recalled my own difficulties in deciding how much the communist world as it actually developed owed to Marxism and how much, adventitiously, to the Russian tradition. After all, it was in Russia that the first communist revolution took place, and communism became a Russian export.

When I read the Marquis de Custine's *La Russie en 1839*, I recognised many things about Nicholas I's Russia that I saw in the communist world. To take one small apercu: he said that the square in front of the Winter Palace was such that a crowd in it would be a revolution. And indeed, in Pyongyang, a crowd that gathered spontaneously in any of the city's vast public places would be a revolution, as it was in Petersburg.

Yet for all that, communism did introduce something new in the world. The communists frequently executed more people in an hour than the Romanovs had in a century; and here it seems reasonable to take seriously one of the three laws of dialectical materialism, namely the law of the transformation of quantity into quality. The communist regime was qualitatively different, and qualitatively worse. It was orders of magnitude more immoral.

Does it matter if Myers gets North Korea wrong (if, that is, I am right)? If one denies that the monstrous regime has any intrinsic connection with Marxism, or with Marxism-Leninism, then the latter is let off the hook. And on the hook, in my opinion, is where it belongs.

32
Of Love, etc.

The government makes me angry, but my wife makes me much angrier (as well as much happier, of course). This is yet another illustration of the truth of Doctor Johnson's dictum that public affairs vex no man, at least not very greatly and not within quite a wide range of government policy. The personal may or may not be political, but it is definitely what concerns us most. Let the heavens fall, so long as we are happy at home.

Paul Hollander's new book, *Extravagant Expectations* (Ivan Dee), is not only about the personal, but about the personals, those small-ads in various publications in which people seek what used to be called a lover, paramour or consort, but must now be called a partner. With the eye of the true sociologist, he uses these brief messages to peer into the soul not only of the individuals concerned, but of western, and particularly American, society. What emerges is both funny and melancholy, and by no means reassuring: but perhaps society has always been one of those things that, like death and the sun, cannot be stared at for too long, for it is never reassuring.

The few words of the personal do not allow individuals easily to distinguish themselves from others of their type, so it is particularly suitable for examining the types that resort to them. Perhaps, indeed, these advertisements would be better designated as the impersonals. It was clever of Professor Hollander to spot this; like Autolycus the Rogue he is a snapper-up of unconsidered trifles, but he is also a master of extracting deep significance from apparently trivial phenomena. He sees a whole world in a grain of sand, while there are all too many of us who do not see a grain of sand even in a whole world.

In modern society, people are supposed to shift for themselves, to develop their own lives according to their own conceptions, and to experience what is known as 'personal growth,' a process incapable of definition that is supposed to last until five minutes before death. People are no longer born into a social role that they are assigned to fill until they die, simply by virtue of having been born in a certain place to certain parents. In theory, at least, every man in modern society is master of his own fate. Where he ends up is a matter of his own choice and merit.

In so far as modern society actually conforms to this ideal, it obviates the frustration of the man of talent who can get nowhere because of a rigid caste system that keeps him in his place, which is to say where he was born. Not only his, but every, career is open to all the talents. The problem with meritocracy, however, even in its purest imaginable form, is that few people are of exceptional merit. The realisation that the fault lies in us, not in our stars, that we are underlings, is a painful one; and in the nature of things, there are more underlings than what I am tempted to call overlings. A meritocracy is therefore fertile ground for mass resentment.

Moreover, in such a society everyone is supposed to find his or her own mate. The age-old system of arranged marriages comes to seem anachronistic and humiliating, a wound to the newly-liberated ego. In actual fact, that system, in which both prospective spouses exercise the right of veto, seems to me an eminently sensible one, somewhat better in practice and more realistic about human nature than the romantic individualism that, at least nominally, reigns supreme in our societies. For example, the parents of a close Indian friend of mine selected six women as possible wives for him; they selected them on the basis of their religion, caste, level of education, knowledge of English, and so forth. The wife he married was actually the fifth of the women whom he met; when he asked her at the formal meeting between her, him and both sets of parents what she was reading, she replied, *Les liaisons dangereuses*. He thought, 'This is the woman for me,' but he nevertheless went to see the sixth – a fact of which his wife of more than thirty years, the fifth woman, reminds him now and then. Few better marriages are known to me.

Our system is founded on our inalienable right to pursue our own happiness in our own way, a right that is supposed actually to result in happiness, or at least in more happiness than if we did not enjoy such a right. When it comes to choosing a mate, we have to consider only our inclinations, and not such things as obligations to society or parents. And when the marriage no longer suits, when the immortal beloved begins to bore us in a way incompatible with the chronic ecstasy we have come to believe is the only worthwhile state for a man (or woman) to be in, we dissolve the marriage and go off in search of another potential source of undiluted and everlasting bliss.

Unfortunately, the pool of candidates has in the meantime contracted. Gone are our student days, when the field seemed so ripe for the harvest. The field has thinned out like the hair on a man's head. It is time to advertise.

The self-presentation of the advertisers depends on the type of publication in which they advertise (increasingly on the internet, of course), though they also select the kind of publication according to their own predilections. The funniest, but in some ways the saddest, of the chapters in this book is about the advertisements in the *New York Review of Books* and in the alumni magazines of great universities. Here, taken at random, is one:

> Savvy, sassy, sweet and really good-looking, with lively intellect and mischievous sense of irony... Slender, willowy, with shoulder-length hair – resemble younger, funnier Susan Sarandon in looks, politics. Fun, empathic, adventurous. Can talk travel, movies, baseball as seamlessly as economics, literature, politics. Adore Clint Eastwood, Stegner, film noir, Picasso, Vermont tomatoes, swimming badly, exquisite discoveries.

The mixture of ordinariness on the one hand (baseball and Clint Eastwood), and preternatural sophistication on the other – no New Hampshire tomatoes for her, she would gag at the very thought of them – is absolutely typical. All the people advertising are Renaissance men and women, capable of appreciating all that the world has to offer; they are both as nature made them and highly cultivated. 'Nature-lover but can do black tie at the drop of a hat.' They throw in their liking for fine Bordeaux wine, or the Umbrian countryside, as markers of their class, but also their liking for the Moody Blues and roast chicken, to demonstrate that they are not snobs.

The question naturally arises as to why such paragons need to resort to these methods to attract a mate. It seems that we have moved from arranged marriage to social isolation, to quote Professor Hollander's succinct phrase. One is inclined to laugh, or at least to smile, at the self-presentation of the advertisers in the *New York Review*, but it surely takes little effort of the imagination to understand the sadness, the human longing, behind that presentation. A world in which elegant, intelligent, highly-educated and probably sensitive people (for surely downright ugly and uneducated people would or could not advertise in this way) are so obviously lonely has something deeply disconcerting about it, and raises awkward questions. Is the individual search for happiness enough of a philosophical foundation for the good life?

Because of the restricted number of words in which they must be couched, the personals are written in a kind of code; and yet, at the same time, one suspects that if the advertisers had an extended space in which to describe themselves, they would be at something of a loss to know what to say, and nothing much more individual would emerge about them. They are individuals without individuality.

In India, where advertising in newspapers for spouses had long been perfectly normal, a technical code has been elaborated, but it attaches to very definite or tangible qualities. The word 'wheaten,' for example, is used of the complexion, and represents a darker shade than the word might appear to imply to the western reader (in India, a dark complexion is no asset, and is even an impediment to marriage that can overcome any number of other advantages, such as a good income). There is a hard-headedness or ruthlessness to the Indian code that is mostly missing from the American personals, which partake of all the specificity of psychobabble.

What is one to make, for example, of the Alabama teacher who says of herself in her advertisement:

> I love hanging out... I am typically up for whatever and have
> fun in most any situation...?

One could sit next to her on a bus and not realise that it was she; for how does what she says about herself distinguish her from anyone else? Clearly, she is less sophisticated than the advertisers in the *New York Review*, and I doubt that she could tell a Burgundy from a Bordeaux, but there are probably ten million young women who could say 'I am up for whatever.'

Whenever I sit on buses or trains I like to listen to the conversations of the people around me. Mostly people talk of themselves, but rarely in such a fashion that you (or their interlocutors) can get any concrete idea of their lives. Their words lack any clear denotation. They are like a confession of having sinned in general, without any details as to the actual occasion when the sins were committed. It is self-exposure without self-revelation.

In addition to the personals, Professor Hollander analyses the books of advice on dating and mating that sell by the million and make their authors very rich. Presumably people do not buy them unless they feel that something is wrong or missing in their lives. Again one is tempted to laugh at the cold-blooded instrumentality of Dr Phil's advice to women about how to catch a man:

> Eye contact is an especially powerful presentation tool... You choose what type of presence you want to radiate in a room... Don't just show up at a social situation. Show up with a plan... Create your sound bite. Explain who you are in twenty words or less. Define four or five things you can talk about at any time to anyone.

People who think of 'eye contact' as a 'presentation tool' are destined for superficial and unsatisfactory relations with other people, all the more frustrating because they are looking to those relations to fill their own inner emptiness, to perfect their highly imperfect and lonely lives. When the presentation tool has worked, the person upon whom it has worked is soon found to be unsatisfactory; it must soon be employed again. Mr or Miss Right never appears.

It is wrong to laugh because the suffering that these manuals of human relations reveal is very real. One has only to imagine someone poring over them with close attention to appreciate this. Who but someone genuinely unhappy and perplexed could believe in 'Your defined product' (Dr Phil) as a possible solution to loneliness?

At the root of the problem is our belief in the perfectibility of life, that it is possible in principle for all desiderata to be satisfied without remainder, and that anything less than perfection, including in relationships, not only is, but ought to be, rejected by us. We cannot accept that we might at some point have to forego the delirium of passion for the consolation of companionship, that Romeo and Juliet is fine as catharsis but not very realistic as a guide to married life at the age of 56. We cannot have it all.

We are in revolt against what Hollander calls 'the limitations imposed by our mortality, genes, social and physical environment, and chance,' as Satan was in revolt against God. *Extravagant Expectations* is an excellent illustration of how the examination of a seemingly minor social phenomenon can soon lead to the deepest questions of human existence.

33
Knowledge Without Knowledge

R ecently I reviewed a short book by David Horowitz, a man who has changed his political and philosophical outlook somewhat down the years, to put it no stronger. He has mellowed with age, a process that seems perfectly normal, indeed almost biological, until one remembers than not everyone does mellow with age. Some remain mired in the swamp of their youthful convictions.

As it happens, I had in my library a book edited in 1971 by Mr Horowitz, in the days when he as still a leader of the American New Left. It was a collection of essays about the life and work of Isaac Deutscher, the British Marxist biographer of Stalin and, most famously, of Trotsky. Deutscher was also a prolific journalist and essayist.

Isaac Deutscher was born in Poland, a subject of the Tsar, in 1907, and died a British citizen in 1967. His move to England in 1939 saved his life; if he had either stayed in Poland or moved to Russia (where he was offered a post at a university) he would almost certainly not have survived the war.

Deutscher was an infant prodigy, brought up as a religious Jew but losing his faith at an early age. He transferred his religious longings at about the age of twenty to the secular faith of Marxism, and never lost that faith to the day he died. Happy the man who lives in his faith, but unhappy the man who lives in a country in which his faith has become an unassailable orthodoxy.

When one reads Deutscher aware of the fact that English was his sixth or seventh language, one is truly astonished, for his prose in his sixth or seventh language is lucid and even elegant, with absolutely no hint that he is not a native-speaker, and a highly-educated one at that. As a sheer linguistic feat this is, if not completely unexampled, very remarkable indeed. Although a Marxist, he modelled himself as a stylist on Gibbon and Macaulay, and if he does not quite reach their level – well, who does nowadays?

His language was clear, but his thought was not. He was what might be called a dialectical equivocator, made dishonest by his early religious vows to Marxism. This made him unable to see or judge things in a common-sense way. His unwavering attachment to his primordial philosophical standpoint, his irrational rationalism, turned him into that most curious (and sometimes dangerous, because intellectually charismatic) figure, the brilliant fool. He was the opposite of Dr Watson who saw but did not observe: he observed, but did not see. He was the archetype of the man, so common among intellectuals, who knows much but understands little.

A good example of this capacity to misunderstand despite a great deal of knowledge occurs in his posthumous short book, *Lenin's Childhood*. When he died, Deutscher was working on a projected biography of Lenin, but only the chapter devoted to Lenin's childhood existed in anything like publishable form; it was edited by his wife and collaborator, Tamara.

From the purely literary point of view, the fragment is characteristically excellent, the very model of its type, written in beautifully balanced prose and with a judicious amount of detail. Of course, an account of so factual a matter as Lenin's childhood must be influenced deeply by the biographer's overall assessment of Lenin's character and achievements, for the child is father to the man and it is the final character and achievements of that man that the childhood in part is to explain or at least prefigure. In Lenin's case, we are interested in the childhood because of what he became, not for its own sake; and it is inevitable that we shall look for different germs of the future in it if we consider Lenin the nearest man to the devil incarnate who has ever existed from those that we shall seek if we regard him (as Deutscher did, according to his wife) as 'the most earthly of all who have lived on this earth of man' – clearly a religious way of putting it, incidentally. What is to be explained differs completely in the two cases: the person who thinks of Lenin as the frozen-blooded murderer who could order executions by the thousand without so much as the flicker of an eyelid will look for different things in his childhood from the person who thinks that he was the brilliant saviour of the world.

Be that as it may, there is a single reference to Dostoyevsky in the fragment that illustrates perfectly Deutscher's learned obtuseness. Writing of Lenin's father, an inspector of schools who was loyal to the Tsar and the Orthodox church, Deutscher says:

> In his young years memories of the suppression of the Decembrist rising were still fresh and forbidding. Then came the terror that crushed the Petrashevsky circle and broke a man of Dostoyevsky's stature.

Admittedly I do not read Russian, unlike Deutscher, but still I do not think it would be possible to write a single sentence that could misunderstand Dostoyevsky more fundamentally, completely and deeply than the second that I have just quoted. Far from breaking Dostoyevsky, his imprisonment, death sentence, reprieve and exile were the making of him, in the sense that they were the experiences upon which his subsequent philosophy, for good or evil, was based.

The reason for Deutscher's most elementary error is obvious. Lenin was the very embodiment of precisely the kind of ruthless, murderous revolutionary to whom Dostoyevsky was drawing attention: he was the very fulfilment of Dostoyevsky's prophecy. Dostoyevsky foresaw not by 'scientific' deduction, *à la* Marx, of course, but rather by intuition and imaginative insight into the souls of men, and he was vastly more accurate as a guide to the future than Marx ever was. But to have admitted this would have been to blow apart Deutscher's whole world-view, the world-view that made his very considerable literary labours meaningful for him, and for which he had, when in Poland, risked his life. So he preferred to see Dostoyevsky not as a man who, as a result of his experiences (in conjunction with native talent, of course) had penetrated to what others had not penetrated, but as a broken reed, a man successfully terrorised by the powers that were. For Deutscher, Dostoyevsky wrote what he did not because he believed it to be true, or had any insight into the nature of things, but because he had been rendered neurotic and cowardly by fear. Nicholas I therefore broke Dostoyevsky, though in truth it would be more true to say that (unintentionally no doubt) he made him.

One of Deutscher's collections of essays, always intensely readable, was called *Heretics and Renegades* (published, of course, by a capitalist outfit – but then, as Lenin said, the capitalists will sell you the very rope with which you can hang them).

The title – from 1955 - is instructive. Four legs good, two legs bad: for Deutscher, the correct slogan was heretics good, renegades bad. It wasn't difficult to see why he should have believed this.

He regarded himself as a heretic but not as a renegade. He was a heretic because he adhered neither to the catholic church of Stalinism, nor to the protestant one of Trotskyism, but rather insisted that he was the one true Marxist, the only other communicant of his church, at least until he was taken up (rather to his surprise and delight) by the students at Berkeley and elsewhere in the United States, and also by the Bertrand Russell Peace Foundation, was his wife, Tamara.

A heretic for him was therefore a hero, he being one of course; but a renegade, the person who had once been a communist but had abjured the faith altogether, was, in Islamic terms, an apostate. The first essay in the book is an extended review of the famous book *The God that Failed*, a collection of six essays by ex-communist intellectuals who explain their renunciation of the faith altogether – for Deutscher renegades all. For them, it was not only that communism failed completely to live up to its ideals, but that its ideals were wrong and therefore intimately and inextricably related to the horrors that followed.

For Deutscher, by contrast, the ideal of a society in which peo-ple were completely undifferentiated by class, in which a spontaneous abundance arose in which people produced for use and not for profit, in which no one exercised more power than any other person, remained not what it always was, an adolescent and not terribly intelligent dream, but real, something directly to be aimed at; and never mind if people initially possessed of this vision (the product, usually, of profound and often unbalanced resentment) had so far killed millions of people. They had merely gone about it the wrong way. Deutscher, the most egocentric of men despite a pretended humility, would show them the right way:

> He [the ex-communist renegade] no longer throws out the
> the dirty water of the Russian revolution to protect the baby;
> he discovers that the baby is a monster than must be stran-
> gled.

The death of tens of millions becomes mere dirty bath-water; the baby – presumably the core of the Soviet Union, its ideal, not its practice

– is still beautiful.

Deutscher reproached the renegades of *The God that Failed* for their tendency to abstraction, of uninterest in concrete realities of the world around them, but you can't get much more abstract than calling mass famines, purges, the gulag, mere dirty water. It is no surprise, perhaps, that a man who can do so has about as much sense of proportion as a young child from whose hand a toy is removed. In his essay, *Post-Stalinist Ferment of Ideas*, Deutscher has this to say:

> Having for decades lived under its own (triumphant!) brand of McCarthyism with its loyalty tests, charges of un-Bolshevik activities, witch-hunts and purges, terroristic suspicion and suspicious terrorism, Soviet society is now driven by self-preservation to try and regain initiative and freedom of decision and action.

The *suggestio falsi* in this is that the Soviet Union was in some way imitating McCarthyism; the *suppressio veri* is that, even taken at its worst (thousands of people dismissed from their jobs, for example), McCarthyism is not to be compared with (say) the forced construction of the White Sea Canal, in which up to 100,000 people died, just one – of many - of the episodes of Soviet *de facto* mass murder. It is difficult not to conclude from the passage I have just quoted that Deutscher was not an unprincipled liar – in defence of his principles.

In his review of Orwell's *Nineteen Eighty-Four*, titled *The Mysticism of Cruelty*, Deutscher says that it 'is in effect not so much a warning as a piercing shriek.' In the course of the essay, he says of the Great Purges in Stalin's Russia:

> To be sure, the events were highly 'irrational;' but he who because of this treats them irrationally is very much like the psychiatrist whose mind becomes unhinged by dwelling too closely with insanity.

To reduce the Great Purges to the status of events, a word that applies to all human happenings whatsoever, is to deny their exceptional or special historical significance, again with the motive of preserving the beautiful, rosy baby of Deutscher's absurd and shallow ideals. Deutscher's use of quotation marks suggests that he thinks the Great Purges were rational, which in a sense they were: that is to say they served the purpose of concentrating Stalin's power, even if the accusations in the purges were themselves absurd and without empirical foundation (not, of course, that the accused men were therefore admirable men, very far from it).

Now in a sense all human desires, in the last resort, are irrational, or rather arational (what cannot by definition be rational cannot by definition be irrational). But to suggest that treating the purges as irrational is itself a sign almost of madness is to accept the purges' *ratio*. Deutscher's objection to murderous purges was really that the wrong people were purged, not to the murderousness.

Deutscher was a fine example of the scholar who knew a lot and understood little (including, or especially, himself). A man may smile and smile and be a villain. A man may read and read, and experience and experience, and understand nothing.

34

Forgiveness Is a Kind of Wild Justice

Recently I was asked at a public discussion of crime and punishment at which I was a speaker whether I thought it was right that the government (in Britain) had made it illegal for an employer to ask a prospective employee whether he had a criminal record and, if so, its nature and extent. This is a question that I have turned over in my mind, or at least let bubble away in my subconscious, ever since, for it in turn raises several interesting and important questions.

Most of the people in the audience, I suspect, thought that the rule was right, for it is both just and merciful (rarely are the two qualities so neatly conjoined) to give criminals who have purged their legal punishment a second chance. The idea of redemption is perhaps a legacy of Christianity even among those who are not Christians themselves. And the notion of forgiveness is especially attractive to people who do not want to appear primitively vengeful. Working as I did in a prison for many years, I often tried to put myself (mentally) in the position of a prisoner leaving prison: where would he go, what would he do, how would he keep himself in a way that did not involve crime?

With regard to the latter question, a couple of statistics are instructive. The prison department in Britain once published the ages at which adult prisoners were received into prison: 97 per cent of those who had committed burglary, and 98 per cent of those who had committed robbery, were between the ages of 21 and 39. This meant, or suggested, that criminality, at least of these two types, ceased spontaneously at the age of 40: assuming, of course, that it did not mean that the burglars and robbers had simply become more adept at crime and therefore evaded detection.

Crime in general is a young man's game; but the fact is that if former criminals can keep themselves after the age of 40 by some legal means of other, they could have done so before the age of 40 also. In other words, their recidivism (for most of the criminals in prison are recidivists and not first-timers) is the result of a lack of will, not a lack of opportunity, even if, as has sometimes been suggested by those who want to ascribe crime to anything other than the decision of the criminal to commit it, the change in their conduct at the age of 40 is ascribable to falling levels of testosterone. In other words, no special efforts are necessary on behalf of prisoners leaving prison, even if nevertheless some such efforts ought to be made: eventually they will do everything for themselves.

But let us return to the questions of justice, mercy and forgiveness, tackling the latter first. The willingness and ability to forgive or overlook is essential to good human relations because we are none of us angels, we all do things we should not, and some of us even have habits irritating to those closest to us (in my case that of never passing a bookshop without buying a book, which my wife finds very irritating). If we did not have the capacity to forgive, every argument would end in divorce or murder, or at any rate in some very unpleasant consequence.

But it does not follow that what is necessary in some circumstances is necessary in all, any more than it follows that a medicine that is good for you in a certain dose must be twice as good for you in double the dose (though I have met patients in my medical career who did believe that, often with near-disastrous consequences).

In order to have the *locus standi* to forgive, the harm that someone does must be done, at the very least in part, to oneself. If someone robs you in the street, I have no right to forgive him; only you have that right. Moreover, even if you do forgive the robber, your forgiveness, morally grand as it might be (though it might just as well be cowardly or pusillanimous), has no claim to determine the treatment of the robber by the law, any more than your vengeful feelings, if you had them, would have done. Revenge, said Bacon, is a kind of wild justice, which the more man's nature runs to, the more ought to weed it out; the same might be said of forgiveness, except perhaps that wild would not be the qualifying word to use of the justice that would result from it. The law is instituted precisely to supersede the effects of incontinent emotion, whether it is of the punitive or sentimental kind.

Forgiveness, then, unlike mercy, has no place in the law. A pardon is not forgiveness, it is an exceptional act which in no way lessens the guilt of the pardoned.

Mercy is an implicit recognition of the imperfection and imperfectability of man, and that it is unreasonably rigorous to expect perfect behaviour of any featherless biped. As Hamlet said, if we were all treated as we deserved none of us would escape a whipping; it does not do, then, to administer justice as if no other virtue or desideratum than justice existed.

Those who think that employers should not have the right to ask applicants for jobs whether they have a criminal record lose sight of these considerations, as well as others. They believe that ex-criminals would find it harder to find employment if employers knew about their past, and that an inability to find work is one of the reasons so many criminals return to crime. But this is to suppose that ex-criminals have superior rights to those of employers who, in order to protect those rights, must blindly accept risks that they might otherwise not be prepared to run.

As it happens, even those who think that employers should not have the right to ask about job applicants' criminal record do not believe this of every kind of crime. They do not believe that schools, for example, should not know anything of a prospective employee's record as a paedophile. However liberal a person may be, there always one corner of his heart reserved for vengefulness towards at least one category of person.

But let us disregard this for a moment in order to conduct a small thought experiment. Let us suppose that you need (or at any rate want) a gardener. There are two applicants, equal as far as you can tell in their gardening abilities, and equal in charm, etc. The only difference between them that you can find is that one of them has a criminal record for stealing from his employers. Which of them do you choose?

Most people, I suppose, would choose the gardener with an unblemished record of honesty. But there are some generous souls who, anxious to do good, might choose the man with the criminal record. We shall not enquire further whether their generosity is moral exhibitionism, the desire of a moth to fly to the flame, or obedience to an abstract Kantian categorical imperative. The fact is, however, that there are people who are willing to overlook a criminal record in order either to do some good to society or to feel well about themselves. This, incidentally, applies as much in the sphere of personal relations as in the field of employment. Murderers, especially the most notorious, seldom lack for offers of friendship or marriage from precisely the same kind of people as their victim or victims.

I remember, for example, the case of a woman in our hospital who had just had her jaw broken by her lover. According to her, he had once 'snapped' her forearm, giving her a fracture by gripping it in his hands and applying force. She had met him not long after he had been released from prison for having killed a former girlfriend.

Leaving aside the question of whether he should ever have been released from prison, her choice of boyfriend seemed to me distinctly inadvisable. I did my best to persuade her of the dangers, and at first she seemed convinced. We closed the hospital ward to him, we found her a safe place to go where he would not be able to find her. But at the very last minute she decided that love was more important than safety, she relented and left the hospital with him, arm-in-arm, laughing and joking with him. No doubt they went straight to a pub where, once he had had a little to drink, he would accuse her of having been unfaithful to him.

The risk she ran was of her own choice: a foolish choice, no doubt, but a choice nonetheless. Now it seems to me that, by contrast, the state has no right to make people run risks which are easily knowable but unknown because it wants to achieve another goal, even a laudable one such as the reintegration of criminals into respectable society.

Let us consider the case of the woman above, under slightly different circumstances. Supposing she had had her jaw broken but did not know that the perpetrator was a convicted murderer recently released from prison, but that I did know this. Would it have been my duty to warn her?

Clearly it would even if, as a matter of statistics, it was far less likely that he would one day kill her than that he would never kill her (for most released murderers do not kill again, even if their murder rate is very high by comparison with those who have never murdered). I would think it was important that she should be in possession of the information in order to make a choice.

Now someone might say that, in order for her to make her choice in a truly informed way, she would not only have to know that the boyfriend was a killer, but what were the statistical chances of a man such as he killing again – otherwise, she might make her choice on the basis of a mere prejudice against murderers. And since prejudice is the basis of all discrimination, it would be better for her to know nothing of this man than to know only that he was a murderer. Murderers have their human rights too, and the avoidance of discrimination on the basis of prejudice is a goal of such overriding importance that allowing people to take unknown but knowable risks is a small price to pay for it.

Such an argument would be absurd, and for the state (as in Britain) to force private individuals or companies to bear risks that they could easily avoid is highly dictatorial.

In fact, no one is free unless he is free to act upon his own prejudices. If he does so act he might be a very unpleasant and bigoted person indeed, or he might be an exceptionally generous and warm-hearted one: it all depends upon what his prejudices actually are. But the idea that the government should determine what prejudices people must not act upon – for example that an extensive criminal record might make a prospective employee less than a safe bet – is totalitarian.

For myself, having worked in a prison for many years, I have a soft spot for criminals – or at least, for some criminals. I am among those who would be inclined personally (and within reason) to give them a chance, if I had jobs at my disposal. I would even be prepared to be disappointed, to find that the thief whom I had found charming and thought wanting to turn over a new leaf had actually stolen from me. I would pat myself on the back because I would think that I had performed a good and charitable act by employing him. But I would also think it the grossest act of tyranny to require my neighbour to behave in precisely the same way. I can take a risk myself that I have no right to demand that others take. It is typical of governments that they should not understand the distinction.

35
The End of Charity

A short while ago in Sao Paulo I witnessed in a restaurant something that moved me. Among the waiters, dressed in the same uniform as the others - that is, white shirt, black trousers and burgundy bow tie – was a young man with Down's Syndrome. He was obviously very happy and proud to work there and to make himself useful: he cleared dishes, wiped glasses, and so forth. (The restaurant, incidentally, was a good one.)

I do not know whether or not the waiter with Down's Syndrome was connected in any way with the owner or manager of the restaurant, but his employment there seemed to me an imaginative and efficient act of management, and not merely a charitable one. Of course, the young man in question benefited – you could see that by the pride on his face. But so did the restaurant as a business, in more ways than one.

The effect on both customers and staff of employing the young man was likely to be highly beneficial. Customers would probably see him and conclude that the owner was a decent and therefore an honest man, not unscrupulous, trustworthy. The presence of someone patently more unfortunate than they would inhibit their inclination, if any, to petty complaint; they would feel ashamed to carp. Satisfaction rushes in where complaint fears to tread.

As for the staff, they, in keeping an eye open for the welfare and safety of the young man, would be aware that they were performing a meritorious social duty and not just helping the owner to a profit; and behaving well reinforces good behaviour. Their propensity to complain, if any, would likewise be reduced. Though strict and narrow analysis might demonstrate that the waiter with Down's Syndrome was not worth his wages – slowness, low productivity, breakages, etc. – his intangible morale-boosting outweighed by far his deficiencies as an employee.

I have noticed this effect before. For example, I have been asked several times to a certain radio studio to give the public the inestimable benefit of my opinion, usually in a few seconds flat. (All opinions should be expressed as concisely as possible, but not more concisely than possible.) And at this certain studio is employed as a receptionist, who shows guests to the various rooms in which they will give vent, a young black woman who is both blind and somewhat physically handicapped, requiring sticks to walk.

From the point of view of Taylorian, time-and-motion efficiency, perhaps, this would seem a foolish arrangement. The person whom she is supposed to be assisting ends up assisting her in the performance of her duty. Is this political correctness gone mad?

No. The young woman has a delightful personality, a cheerful disposition, an evident liking for the public with whom she has to deal. Once again, factious complaint about trifling inconvenience – being kept waiting a few minutes, for example - is rendered not only shameful but absurd. For what are a few minutes' wait to set against a lifetime of blindness and a deformity that makes each step an effort of will? One would have to be a swine to complain in her presence (not that such swine cannot be found, of course).

On the way to the recording room, the receptionist asks the guest whether he would mind helping her with the various doors en route. By the time he sits down before the microphone, therefore, he is in a thoroughly good or mellow mood, aware of what a nice fellow he is for having helped a poor unfortunate with such good grace: though it is in fact she who has helped him.

When, then, someone else says something foolish, preposterous or even nasty on air, the guest feels no anger, and replies as if a soft answer not only turned away wrath but convinced the foolish. Indeed, if the young woman should ever be sacked from her job, it will be because of her calming effect, because broadcasters increasingly demand (at least in Britain) that there should be confrontation rather than discussion, the former – supposedly - being infinitely more entertaining than the latter. Broadcast confrontations are now to the British what gladiatorial combat was to the Romans.

I shall give just one more example of the salutary presence of the obviously handicapped (if you give too many examples you become boring, if you give few you are accused of being anecdotal). One day a young mentally handicapped man was admitted to the prison in which I worked as a doctor, on a charge of having sexually molested a young woman. The assault was alarming to her rather than dangerous; the young man was of such a physique that it was difficult to imagine him overpowering anyone, though he had not even tried to do so.

He was sent to prison not because it was the right place for him, but because it was the only place that could be found for him at the time: a flurry of humanitarianism having previously closed down all the other institutions that might have cared for him.

Normally sex offenders, of whatever kind, are regarded by other prisoners as the lowest of the low: it is only thus that prisoners can boost their own morale by conceiving of people worse than themselves. Indeed, sex offenders have generally to be separated for their own safety from other prisoners; if not, they are attacked and injured, sometimes seriously. Moral depravity is not incompatible with moral indignation – and, of course, vice versa.

On this occasion, however, the prisoners recognised that the young man was worthier of pity than indignation: they made an exception in his case. Indeed, they looked after him with solicitude because he was so obviously distressed by a situation that he could not understand. (I shall not easily forget his howls of distress.)

They comforted him, they shared their things with him; they succeeded in calming him down. And while he remained in the prison, the prisoners in his location behaved better; they were softened by the licence his presence gave them to relax their hardness, their toughness, their cynicism that was otherwise essential to survival in the social, or anti-social, world of prison. Normally, such a lowering of the guard brings instant retribution; but not in this case.

Now the politico-bureaucratic soul, when it becomes apprised of such cases, immediately perceives in them an opportunity for increasing its own dominion. For if it is the case that the presence of the handicapped often improves morale, and even morality, is it not obvious that such a presence should be made obligatory, decreed by law? Will this not make the occasional the general, and thus add to the happiness of mankind? After all, not every manager is as charitable and imaginative as those of the restaurant or radio studio, nor does every prison location have its handicapped person to soften the mores of the other prisoners. Acts of charity, understanding and solidarity must therefore be legislated for, so that they become institutionalised. For what is possible on one occasion must be possible on all occasions.

Justice and equity demand it. Why, for example, should the young man in the Sao Paulo restaurant have benefited from the enlightenment of the manager or owner when others, perhaps thousands of others, had no such lucky chance? Worse still, it is likely, even probable, that the young man in question benefited from some personal connection from which, by definition, others, just as deserving as he, could not benefit. And nothing – nothing – is worse than injustice.

Where equity is concerned, it would be far easier to insist that the young man in the restaurant be made redundant than that other such young men (and women) be employed. There could then be no accusation of unfairness or injustice, since all in his position would be treated alike. But such a solution to the problem would provide no opportunity, or very little opportunity, for the politico-bureaucratic class to intervene in the affairs of men. Much better, and more obvious, from that class's point of view, would be legislation to compel what was previously only voluntary. Indeed, a decree that every enterprise should take on handicapped staff offers a rich field for inspection and bureaucratic bullying in the name of humanity.

How delightful are the prospects! Needless to say, what constitutes handicap is a matter of endless possible dispute, because handicap is not categorical, it is dimensional. What is a serious handicap for a footballer is not necessarily an serious handicap for an accountant, and vice versa.

Many delightful, intractable and therefore profitable disputes loom. Is drug-addiction, for example, a medical condition, and therefore a handicap? Should enterprises therefore be obliged to take as employees the percentage of drug-addicts that exist in the wider world? And how wide should that wider world be? If an enterprise happens to be sited in a community in which there are no drug addicts to employ, should it go looking for them?

But the fundamental objection to institutionalising charitable acts by government fiat is that it hardens the heart and makes compassion almost impossible: there can be cruelty without discretion, but not compassion or real feeling (this is not quite true, but almost true).

Let me illustrate what I mean by reference to the social workers in the hospital in which I once worked. I take it as being beyond reasonable doubt that there are some people who fall on hard times through no fault of their own and who are therefore particularly deserving of assistance. But I was unable to persuade many of the social workers in my hospital of this, because if some cases were particularly deserving of assistance it followed that others were not; and it was part of the ideology of the social workers that they should not assist people according to their desert, but only according to their need.

This had the horrible consequence that the social workers were not able to exert themselves in proportion to a person's desert; and since the worst-behaved were adept at manufacturing need, it meant that the most deserving were often comparatively neglected. Moreover, this took its toll on the social workers themselves, for with rare and saintly exceptions, it is impossible for people to feel compassion to all equally, irrespective of desert. In other words, the social workers had to suppress their natural feelings; and when such feelings are suppressed long enough, they atrophy and cease to exist. And that is precisely what I observed among them.

It is true that judgments of desert vary, and even where it is agreed as to what constitutes desert error is possible and indeed inevitable: the deserving might be taken for the undeserving, and vice versa. But the consequences of making no judgments are worse than the consequences of sometimes making the wrong ones: indeed, refraining from making a judgment is itself to make a judgment of a kind.

What is given as of right is harmful alike to the donor and the recipient. It shrivels the donor's heart and turns kindness into an unwanted obligation; it renders the recipient incapable of gratitude, to such an extent that he might not even realise that he has received anything (the rioters in London, for example, said they had nothing, when those of them who had never worked or been net taxpayers had never gone hungry, never lacked for clothes or shelter, were provided with electronic gadgets, were guaranteed free healthcare and had received a free education – for them, this was nothing because it all came as of right).

That judgments in the past were harsh or unfeeling is, alas, the case. But that is a reason for refining our judgment, not for refraining from exercising it at all. If we do that, we shall end up with a society of cold comfort, where the faculty of kindness will wither, and where the expression of human solidarity will be confined to paying taxes, an indefinitely large proportion of which will never even reach their supposed beneficiaries.

36

All's Fair in Politics and Celebrity

I f, as the French historian, Pierre Nora, recently put it in a news-paper article, the whole of human history is a crime against humanity, how is one to assess the significance of a single criminal act? And yet the human mind is so framed that it is inclined to see in such a single act all the deceit, evil and delight in cruelty of which Man is ca-pable. One death, said Stalin, is a tragedy; a million is a statistic.

The report of a single dreadful crime is enough to plunge one into despair about the possibilities of human nature. For example, I read recently in a British newspaper a report of a man who picked up an achondroplastic dwarf in a pub and slammed him down on the ground so violently that the dwarf is now paralysed from the waist down and will spend the rest of his life in a wheelchair. Whether or not the final injury was intended by the assailant (was it worse if he did or he didn't intend it?), the act was of insensitivity so gross that it makes one shud-der. Of what would such an assailant not be capable? How is it possible for a human being even to conceive of such conduct, for the thought of it to cross his mind, let alone for him not to know that it was inexcus-ably wrong?

But perhaps we have become so accustomed of late to shifting the boundaries of the excusable and inexcusable, in a kind of amoeboid movement such that, while something is brought into the protoplasm of the excusable, something else is extruded from it, that we no longer really believe in the categories of the excusable and inexcusable at all. Too swift a change in boundaries leads to cynicism about the very notion of boundaries.

Certain character traits are more compatible with a world without boundaries than others: and these are precisely the traits that the British people are fast making their own. Reticence, for example, is now condemned by them as dishonesty and treachery to the self, an unhealthy psychological repression that is incompatible with freedom, and that leads to personal disaster in the end. A kind of incontinent frankness about everything is now admired, even if it is combined with an increasing resort in public life to euphemisms and stock phrases without definite denotation.

Perhaps our loss of belief that there should be boundaries explains why the criticisms of the recent film about Mrs Thatcher, *Iron Lady*, were so beside the point. The real criticism of this film is very simple: that it should not have been made. Whether it was well or badly acted, accurate or inaccurate, plausible or implausible, dramatic or dull, is therefore irrelevant. That so few people saw this fundamental criticism that rendered all others supererogatory is itself a symptom of just how many boundaries we have removed. We are like those who are blind to the fact that smashing dwarves onto pub floors is simply inexcusable.

That a film should never have been made does not prevent it from having excellencies or other defects. Among the former, as has been universally acknowledged, is the performance of Meryl Streep as Margaret Thatcher, a real *tour de force*. She convinces not only as the woman in her prime, but in her decline also; I think (without knowing anything of how the film was actually made) that Meryl Streep must have spent quite a long time observing – and how intelligently! – the behaviour of old ladies in nursing homes. The accuracy of her acting is both remarkable and admirable. My wife, a doctor who specialises in the treatment of the old, felt as she watched that she was back at work.

There is no film about Mrs Thatcher that could be made that would not be criticised as being politically too sympathetic or too unsympathetic towards her; and it is very unlikely that any could be made either that would not also be criticised on the grounds of some historical inaccuracy or other.

For example, it has been objected that *Iron Lady* depicts Dennis Thatcher, falsely, as having been resentful of his wife's political ambition, to which she was prepared to sacrifice the wellbeing and happiness of her family. This is certainly not my interpretation of the way in which the relationship between Mr and Mrs Thatcher is portrayed by the film. It is true that when, in the film, Mrs Thatcher announces to Dennis her intention of running for the leadership of the Conservative Party he accuses her of putting her political ambition first, etc. But, supportive of her as he undoubtedly was, it is surely implausible to suppose that he never ever, in all the years of her political career, uttered such a thought. The outburst does not weaken, but rather strengthens, the impression of the man that is given: by and large humorous, tolerant, charming, modest and dignified, which was certainly my impression of him on the occasion or two on which I met him. It no doubt helped that, in his own sphere, that of business, he had been a successful man; and the film manages a task that is much more difficult to achieve than that of portraying an unhappy marriage, namely the portrayal of a happy one without resort to mere sentimentality.

The political criticisms of the film depend, of course, on your view of Mrs Thatcher. Some critics thought that the political and economic problems that she faced were not sufficiently laid out; but this, it seems to me, is equivalent to the demand that a performance of Macbeth should be preceded by a disquisition on the society of mediaeval Scotland.

It is only natural that anyone who has lived through a recent historical period portrayed in a film should have his own version of it which will not exactly coincide with the film's. Among other errors, *The Iron Lady* portrays Britain as being class-ridden and therefore hidebound, a crude and widespread, and therefore influential, mistake. A class society can be a perfectly open one. If the Tory party had been quite as it is depicted in the film before the ascent of Mrs Thatcher, her ascent would have been inexplicable; and while her ascent was undoubtedly a reward for her enormous determination and strength of character, it is important to remember that, in the Twentieth Century, she was far from unique. Lloyd George, Ramsay Macdonald, Harold Wilson, James Callaghan and John Major were Prime Ministers all of modest or even humble background (their attainments are another question). Indeed, it is even possible that, in the name of egalitarianism, we are transforming a class into a caste society.

A strength of the film (in my opinion) is that it emphasises Mrs Thatcher's early experiences in the Lincolnshire town of Grantham, famous for its association with Sir Isaac Newton and Britain's first serial-killing hospital nurse. Unfortunately these experiences, while formative, served her for ill as well as good, because they remained too present in her mind.

She thought of the British people as she remembered them from Grantham in 1938: honest, thrifty, responsible, self-reliant and longing for freedom. But decades of decline, the welfare state and easy credit have changed all that; very large numbers of them, far from wanting to stand on their own two feet, are only too anxious to stand on the feet of others. They do not fear debt, they fear the withdrawal of credit. Self-respect among them is the dodo of their virtues, dead a long time since; they would rather consume at someone else's expense than not consume at all. Not realising that these changes in the national character had come about, Mrs Thatcher expected more of economic reform than it could ever deliver; she therefore gave the impression of being an economic determinist, a Marxist through the looking glass, as it were.

She destroyed the power of the unions, all right, a very necessary thing to be done, but she left everything else more or less untouched, except in one catastrophic way: she turned managers in the public sector into pretend-businessmen, with all the perks of private businessmen but none of the discipline that the market place exerts, at least on small businesses. Having an almost mystical faith in the science of management, she created what had not existed before, a self-conscious and morally corrupt *Nomenklatura* class of the public sector, a class that her successor-but-one, Anthony Blair, quickly and cunningly made his own. We are living with the consequences still.

This does not emerge in the film, but that is no criticism of it, or at least not a very strong one. First, so subtle a development, while very important (and, indeed, the most important aspect of Mrs Thatcher's domestic legacy), is not easy to convey dramatically; and second. you can't squeeze everything into so small a compass as a film. The film does, however, convey with considerable skill those of Mrs Thatcher's qualities that make most contemporary politicians seem like careerist pygmies by comparison: her courage, her conviction, her devotion to duty. It also suggests, plausibly, that these virtues run riot were the cause of her downfall. Power exercised successfully and for any length of time eventually clouds the judgment. But no one with a half-open mind would emerge from this film *despising* Mrs Thatcher.

The wrongness of the film lies elsewhere: in its depiction of Margaret Thatcher's dementia, for which there is neither artistic nor historical justification. It is intrusive and prurient and nothing else.

It is not pretended that she was suffering from dementia at the time she stood down from office; and, of course, she is still alive. Had she died, however, there is no reason why a film about her career would have been told through the hallucinatory memories during her state of decline. It is the fact that she is still alive that gives the artistic device its spice, if I may so put it; it is of interest only *because* she is still alive.

If her condition is as depicted in this film, she could not have given her consent to it (advance consent is no consent, in my opinion). If, on the other hand, she is not as depicted in it, it is a gratuitous piece of fiction.

It is cruel, degrading and unseemly to exhibit to the idle gaze of millions of strangers (as the makers of the film must hope) a famously self-controlled woman, who took particular and almost fierce care of her appearance in public, grovelling on the floor in a blue flannel dressing-gown, in the grip of a degenerative disease. This is not *Richard II*: it is *Hello!* Magazine with the tact removed.

It cannot be said in the film's defence that it helps to spread awareness of a fell disease. Even if this were so, it still would not justify it: the end does not justify the means, at least not in this case, because (among other reasons) there are so many other means available. Neither is the film a morality tale, because Margaret Thatcher's dementia is not a reward or natural consequence of any of her political actions. The film is therefore merely exploitative and salacious.

By failing to recognise that there are limits to what it is proper for the public to know, and for it to interest itself in, the film adds its mite to the deterioration of public life, not only in Britain, but wherever it is seen. For if public figures are to be treated as if they had no right to private lives, even while they are still living, it is hardly any wonder if only the shallowest, exhibitionist and avid for power go in for public life. And exhibitionism is not honesty, it is coarseness. As it happens, the opening scene of the film depicts the coarseness of modern Britain extremely well. Margaret Thatcher, demented, has escaped from her gilded-cage surveillance for a moment and gone to a convenience store to buy herself a bottle of milk. There, two young men treat her with none of the respect due to age, and push her aside unceremoniously. *They* would not have seen anything wrong in exploiting Margaret Thatcher's state of dementia for the purposes of entertainment: because, of course, in Britain there are no higher purposes than entertainment.

37

To Judge By Appearances

When I was young I wanted to be a bohemian when I grew up. I cannot quite recall how and why I formed this ambition. I suppose bohemianism seemed to be both a way of asserting my individuality (it did not occur to me that bohemians were just as much a herd as an other) and of doing God's glorious work, which was annoying the grown-ups.

I suppose my model was Dylan Thomas, the Welsh poet. He, it seemed to me, had lived as a free man ought to live. This conclusion could only have been drawn by someone who actually knew rather little about his life; and certainly I never had a vocation for excessive drinking. When I reached the age at which I was free to drink as much as I liked, or had money for, I soon discovered that I did not really like the feeling of drunkenness; particularly disagreeable was the sensation when one went to bed that the ceiling above was going round and round. I was fortunate enough also to suffer severely from hangovers, and since (quite apart from the unpleasantness of the hangover itself) I have always been attached to clarity of mind, in so far as I have been able to achieve it, I abjured drunkenness, at least in any regular form. I did not, however, foreswear alcohol altogether; and now not a day goes past, at least unless I happen to be in a place like Somalia, when I do not drink – in moderation.

Dylan Thomas' life (minus the drink, if such a thing can be imagined) seemed a model. At the time, I would not have understood John Malcolm Brinnin's assertion (in his book, *Dylan in America*) that his life was just one plain boring, meaningless, pointless, sordid and avoidable crisis following another meaningless, pointless, sordid and avoidable crisis. The important thing in life was to cock a snook, almost irrespective of the target.

But Dylan Thomas was the real thing, a man of talent if not of genius. Perhaps there is something too theatrical for modern tastes about his public reading of his poems, but I at least am still moved by it. The emotion in his voice and in his lines is real, not bogus; and if his life sometimes seemed almost a caricature of itself, this at least was genuine. I do not see how anyone with even the most minimal feeling for poetry could fail to be stirred, for example, by his *In My Craft or Sullen Art*. Fifty-eight years after his death, his frequently disgraceful behaviour seems a small thing to set against the achievement.

Incidentally, and *a propos* of nothing, his grave in Laugharne church cemetery, in South-West Wales, is one of the most moving graves known to me. It consists of a mound with a simple white cross, painted (and no doubt regularly re-painted) with his name and dates. Beside it, in exactly the same form, is the grave of his wife, Caitlin, with whom he had a passionate cat-and-dog relationship. 'Reunited' does not seem so ridiculous a cliché here, though Caitlin survived her husband by forty years. Perhaps it is partly because I love the landscape of that part of the country so much that I can return to Laugharne cemetery and know that I shall be overtaken by pleasantly melancholic sorrow.

I did not come from a bohemian family, far from it. My mother went to the theatre often, including to all the supposedly shocking new plays, but this was mere diversion from her own unhappiness. I had one bohemian cousin, who lived in Paris for a time, moved among poets, wrote a little poetry, and had a brief affair with Richard Wright (of *Native Son* fame); but she did not have, nor was she allowed to have, much influence on my life.

Alas, as I grew up the times became less and less propitious for real bohemianism. There were a number of reasons for this.

The first was economic: cheap garrets and boarding houses disappeared. Increasing wealth, luxury and housing regulation meant that no one any longer could or was permitted to live in a single room without proper heating, lighting or plumbing. The areas in which bohemians had once gathered were either gentrified, or – to indulge in neologism – millionairified. It is difficult, not much fun, and possibly even slightly dangerous, to be a bohemian in a dingy lifeless suburb without real bars, full of people with nine-to-five jobs trudging to and from work every day. The effect of the demise of the boarding house upon English literature has never been fully documented, but I suspect that it was devastating. No doubt boarding houses with their imperious, mean-spirited, prurient, intolerant landladies had their disadvantages from the point of view of raw physical comfort; but they relieved countless people from the sapping tedium of looking after themselves. They allowed people the greatest luxury of all, the one that we have forgotten: time.

But while real bohemianism has become difficult or impossible – it has gone the way of genteel poverty which, alas, no longer exists or is possible, if only because rents are now too high in the areas where the genteelly poor once gathered – a kind of bogus bohemianism has become the rule. In a sense, everyone is a bohemian now.

Walk down any street in the western world and try to estimate the proportion of people dressed in a conspicuously bourgeois manner. Except possibly in the financial centres (and Swiss cities) it will be very low. The great majority of people whom you pass in the street will be dressed in a manner which, sixty years ago, would have been thought bohemian. And this is so despite the fact that one of the principal pasttimes of enormous numbers of people is shopping for clothes. It is as if they are studiously sloppy.

Naturally, sloppiness in dress is a convention like any other: for if there is one thing that human beings cannot escape, other than death and taxes, it is convention. So the question is not whether a certain form of behaviour is conventional, but whether it represents a good convention.

I am glad to be able to report that, on the question of dress I have changed my mind completely. I am glad to be able to report it because it demonstrates, to myself if to no others, that I am not totally inflexible mentally. Evidence, reason and reflection can still cause me – occasionally – to change my mind, even if it takes me many years to do so.

I was once much struck by a little essay by the novelist Arnold Bennett called 'Clothes and Men'. I will quote from it:

> I would sooner see a fop in the street than a man whose suit ought obviously to have been sold or burnt last year but one. The fop has at least achieved something and is not an eyesore. The scarecrow is an eyesore and has simply left something undone, either from conceit or from sloth. The fop is

not without his use in society. He keeps tailors alert. He sets the pace. He may often be an ass, but he is also an idealist, a searcher after perfection; we have none too many searchers after perfection, and an ass engaged in that quest is entitled to some of our esteem.

Bennett says a little later:

The sole purpose of clothes – whatever it may once have been - is no longer merely to give protection. An important purpose of clothes is to make a pleasing visual impression – partly on oneself but chiefly on other people.

These are not deep or original thoughts, but Bennett is not proud and is ready to answer that criticism:

Platitudes, you will say. They are; but it is astonishing how the most obvious platitudes are ignored by seemingly sensible persons in daily life.

But platitudes are relative to their time; what is obvious in one epoch is no longer obvious, or even true, in another. Let us take one of Bennett's statements, that he regards as a platitude, that an important purpose of clothes is to make a pleasing visual impression, and examine it.

The statement remains only partly true. The part that is true is that an important purpose of clothes is to make an impression, but it is no longer true that the impression that it is their purpose to make is a pleasing visual one, very much to the contrary. The impression that the bohemianisation of dress is intended to make is that the wearer is such an individual, whose real inner me is so unique and valuable, that it is quite unnecessary for him to make any effort to cover it in the rags of mere outer smartness. Do not judge a book by its cover, this form of dress proclaims, or almost shouts; inside me there is the *Summa theologica*.

Actually, it is usually quite possible to judge a book by its cover, at least as a first approximation. It is ridiculous to say that you cannot distinguish between a scholarly work and an airport novel of soft pornography merely by looking at their respective covers. Occasionally you will be wrong in your judgment, of course, and it is always necessary to keep an open mind; but not a mind that is open in the sense that a dam that has collapsed is open.

The sloppiness of modern dress is not the consequence of economic exigency (Bennett in his essay deals with that objection pretty smartly). The late Mr Jobs did not look a mess because he had no money, but because he wanted to look a mess. Perhaps one of his thoughts was that, if he looked sufficiently a mess, no one would object to his having so much money. He would demonstrate thereby that, notwithstanding his great fortune, he was one of the people.

If you look at pictures of crowds in Edwardian times, you will see that practically no one appears in public in a dishevelled state: and this is not because everyone in the picture is well-off. This tradition continued into the 1950s. Even bohemians were not badly dressed: they were differently-dressed. The only people who are in rags are the destitute; they do not want to be in rags.

Our current way of dressing is a sign of our egotism, of our habit of living in a kind of portable solipsistic bubble that goes everywhere with us, like a shadow. 'I am not going to make an effort just for you,' proclaim our clothes. On the contrary, my life is so full of importance, so beyond the right of anyone else to have a say in it, that I shall just put on the first crumpled apparel that comes to hand as a matter of principle.

This, I readily confess, is a revolution in my own thought. When I was young I affected to believe that to look a mess was a sign of inner profundity. I made the mistake of supposing that if the wise are bearded, the bearded are wise. If I adopted the supposed manner of an artist, therefore, I too would be an artist, or at least artistic. And, conveniently enough, the sloppiness of inner profundity required no effort on my part. As everyone knows, a lot of human history consists of man's attempt to escape effort.

Like all virtues, attention to dress for the sake of others can go too far and become the most absurd vanity (though there are far worse and more destructive types of vanity, not least intellectual and moral vanity). The dapperness of Hercule Poirot is ridiculous, and one would not want everyone to be like him. But while his eccentric attention to his own appearance is endearing, the opposite end of the spectrum – complete indifference to the point of dirtiness – is not at all endearing, but repellent rather.

If, from the social point of view, the happy medium is best, it does not follow that both extremes are equally bad.

38

It's a Riot

When the riots in England that astonished the world (but not me) broke out, I happened to be in Brazil. Thanks to the demand for my opinion from around the world – but not from England – I am glad to say that I benefited economically more from the riots than the most assiduous looter. It is truly an ill-wind that blows nobody any good.

After the riots were over, the government appointed a commission to enquire into their causes. The members of this commission were appointed by all three major political parties, and it required no great powers of prediction to know what they would find: lack of opportunity, dissatisfaction with the police, bla-bla-bla.

Official enquiries these days do not impress me, certainly not by comparison with those of our Victorian forefathers. No one who reads the Blue Books of Victorian Britain, for example, can fail to be impressed by the sheer intellectual honesty of them, their complete absence of any attempt to disguise an often appalling reality by means of euphemistic language, and their diligence in collecting the most disturbing information. (Marx himself paid tribute to the compilers of these reports.)

I was once asked to join an enquiry myself. It was into an unusual spate of disasters in a hospital. It was clear to me that, although they had all been caused differently, there was an underlying unity to them: they were all caused by the laziness or stupidity of the staff, or both. By the time the report was written, however (and not by me), my findings were so wrapped in opaque verbiage that they were quite invisible. You could have read the report without realising that the staff of the hospital had been lazy and stupid; in fact, the report would have left you none the wiser as to what had actually happened, and therefore what to do to ensure that it never happened again. The purpose of the report was not, as I had naïvely supposed, to find the truth and express it clearly, but to deflect curiosity and forestall criticism by outsiders.

I was also once asked by the editor of a magazine to read, and then write commentary on, three government reports – one into the origin of a war, one into the origins of some riots (minor by comparison with those of 2011), and one into an outbreak of hysteria about child abuse in a certain corner of England. The findings were more or less interchangeable, though you might have supposed that the three episodes were very different. The cause of all human ills, it seemed, was failure to communicate: and so I began to see what E. M. Forster meant when he finished *Howard's End* with the enigmatic command 'Only connect.' Here was a congenial message for the Third Age – that of psychobabble. If only we connected, that is to say communicated, all would be well, all conflicts resolved. If only the lion would talk to the lamb (and vice versa), they could lie down together; carnivores would henceforth nibble grass.

This is a good moment to return to our sheep (as the French say): or perhaps I should say to our rioters. The commission, according to the headlines in the *Guardian* newspaper, found that 'people needed a stake in society': with the implication that they did not at present have one.

The mental world in which the commission existed was one in which people have a grievance if they think they have one; and furthermore that the grievance about which they feel aggrieved must be precisely what they say it is, failure by others to accept which would be yet another legitimate cause of grievance.

In this mental world, anger and outrage are self-justifying and indeed evidence in themselves of irreproachable righteousness, the main if not the only source of moral authority. I remember a little article in the same newspaper a few years ago about the practice of 'outing,' that is to say the revelation by homosexual activists of the closet homosexuality of certain public figures, whether or not those figures themselves wanted this known. In other words, the activists believed that the public figures involved had no right to privacy but rather had an inalienable duty to bare their souls in public.

The article was written in a for-and-against fashion, giving both sides a fair opportunity to put their case. And the case for the practice was that it allowed people to express their anger, whose object was not specified. In other words it was their anger which made them and their actions morally right; presumably, therefore, the angrier they, or anyone else, felt, the more righteous they became. This does not seem to me to be a recipe for psychic, let alone, social, harmony, but rather for a permanent Balkan war of the soul.

In line with the notion that people need 'a stake in society' in order to refrain from breaking shop windows and taking what they think they have been wrongfully denied (interestingly, the bookshop was the only shop in a very badly looted commercial street that went completely unscathed during the riots), a man called Earl Jenkins – 'who was one of up to 60 youth workers who went on to the streets of Toxteth [a poor area of Liverpool] during the disturbances to persuade youngsters not to get involved' – was reported in the *Guardian* to have said, 'If you've got nothing to lose, you'll do what you can to survive, won't you?'

There was no comment in the newspaper on the deep contradiction in the attitude of Earl Jenkins (let us leave aside the question of how many 'youth workers' in Toxteth are needed to prevent a riot there). For if it is true that the riots were a survival mechanism, why was Earl Jenkins trying to persuade young people not to join in? Did he not want them to survive? Suffice it to say that the objects looted during the riots were not such as people on the verge of famine, or who fear that famine is around the corner, might be expected to loot. They were, rather, the things that spoilt children might be expected to want for their birthday.

The term 'If you've got nothing to lose' in this context is ambiguous. It might mean such penury, such drastic poverty, that you possessed nothing that could have been removed from you. But it clearly cannot mean this, since all the rioters were at liberty, and were clothed, fed, housed, educated (if unsuccessfully), provided with medical care, and given at least a small income, much of which could, in theory at any rate, be removed from them. They could be made homeless; their central heating could be turned off; they could go hungry and literally penniless, made to wear rags; their telephones could be taken from them; they could be deprived of their liberty and even enslaved.

But none of this was going to happen to them and they knew it perfectly well; so in this sense it was indeed true that they had nothing to lose. One of the commissioners appointed to enquire into the riots actually put it succinctly:

When people don't feel they have a reason to stay out of trouble, the consequences for communities can be devastating...

But the reason they 'don't feel a reason to stay out of trouble' is not because they have nothing to lose in the sense of being so deeply impoverished that they have nothing removable from them, it is because they have nothing to lose because they know that whatever they have will never be removed from them, under any circumstances whatever.

Here it is instructive to look at the statistics for house burglary in England and Wales. 750-800,000 such burglaries were known to the police in 2006; the police found the burglars in about 66,000 cases. (The figures for the number of burglaries are underestimated, while those for the numbers of burglaries solved are overestimated, both for technical reasons not necessary to go into, and that we can for the sake of argument ignore.) In that year, just over 6000 burglars received prison sentences. In other words, even if caught, a burglar in England and Wales is not likely to go to prison; but he is even less likely to be caught in the first place. In this sense, then, criminals do indeed have nothing to lose, and possibly much to gain by criminality.

The mystery, then, is not that there should have been riots, but that for most of the time there are no riots. This is a tribute to the inherent goodness, or perhaps to the laziness and cowardice, of man.

The commission's report recommended that 'every child should be able to read and write at an age-appropriate level by the time they leave primary and then secondary school.' Amen to that rather unambitious goal; but asking the question as to why 20 per cent of British youth leave school unable to read and write at an adult level after eleven years of compulsory attendance, and at a cost to the taxpayer of $80-90,000 per head, might have led the commission to a more interesting train of thought about the nature of the British state. How has it achieved this miraculous combination of expense and total failure?

When the commission referred to the 'lack of opportunities for young people,' it might usefully have asked why it was that Britain had had high levels of youth unemployment for many years while simultaneously importing very large numbers of young people from abroad to perform unskilled work. This is an awkward question to ask because it could so easily inflame insensate xenophobia, but it is nevertheless an important one that I have never seen asked in the public prints. By not asking it, we avoid the corollary questions of what social and economic policies have led to this anomaly. And these questions in turn might undermine our confidence in the presumptions of our social and economic policies of the last three quarters of a century. Better, then, not to notice the anomaly, let alone try to think about how it has arisen, and to pretend, rather, that more of the same, perhaps slightly better-refined or targeted (more training for youth workers in Toxteth, for example), will solve our problems.

Some of the recommendations of the commission make the heart sink. It wants children to be protected from excessive marketing, which they believe is an important cause of their indiscriminate materialism and ascription of undue importance to the possession of expensive brands of goods. And in order that they should be thus protected, it recommends the appointment of 'an independent champion to manage a dialogue between government and big brands' – no doubt at a big salary, with a staff of underlings. There is no situation that our new *Nomenklatura* class cannot turn to its advantage; and no end to the number of bureaucracies it can create in order to employ itself.

It is true, however, that a combination of consumerism and utter economic dependence on the state is, like the lot of the policeman, not a happy one. The dependence is (admittedly at some remove) a corollary of the theory of entitlement, and a belief in one's own entitlement is a belief as destructive of the human personality as it is possible to envisage. It precludes gratitude for what one has, encourages resentment over what one does not have, and discourages personal effort except to obtain things at other people's expense. At the same time consumerism, by offering the mirage of personal fulfilment through the possession of trifles, lends an urgency to possession that it might not otherwise have, thus adding to or catalysing the resentments of entitlement. I might add that in a world in which income is in essence pocket money (everything else having been taken care of, albeit at a level less than that desired) consumer choice becomes the only choice that is ever exercised, and thus the model for the whole of human life.

The rioters, then, were (and still are, of course) victims, not of injustice or poverty, but of bad ideas and a rotten culture that, alas, have become truly their own. And the first idea they ought to be disabused of is that there is someone who is either able or willing to come to their rescue.

39
Fairly Just

R ecently, the first of my contemporaries died. She was a student
of languages when I was a medical student. My fellow medical
students and I diagnosed the chronic illness from which she was to die
more than forty years later, the last fifteen of them too incapacitated to
do the teaching work that she loved.

Alas, I had not really kept in touch with her as I should have done –
and now, of course, it is too late. (I am reminded of a wonderful sentence
in a novel by Marguerite Duras: Very early in my life, it was too late.) But
when I knew her well, she was a sweet girl, not innocent exactly - she
was too intelligent for that - but disinclined by natural goodness to think
ill of anyone. I hope that she had her just reward for her charity of spirit.

I shall not descant on mine own mortality, as Richard III descant-
ed on his own deformity: it would be clichéed to do so. In any case,
though I know myself to be mortal, I still live as if I were immortal. My
knowledge of my own mortality is of the same type, and of the same
emotional weight (at least for now) as my knowledge of the date of the
Treaty of Nerchinsk or of Kutchuk Kainardji. I do not feel it in my bones,
but only think it in my mind.

The forty-year illness of my contemporary was a cruelly arbitrary
one, in the sense that its cause was completely unknown; and if any action or habit of hers contributed to its development or progression, it
was quite without her or anyone else's knowledge. By now, every smoker

218

knows the risk he runs (this has been so for generations), and while it cannot be said that he actually deserves what he gets, yet what he gets – if he gets anything at all, for of course there are smokers who suffer nothing that they would not have suffered anyway, the difference between fates of smokers being another manifestation of the purely contingent – is not quite arbitrary. The smoker who contracts cancer of the lung can be said to have contributed to his own downfall, although whether the knowledge of this is a consolation or an additional burden to bear I do not know. According to the modern glorification of the victim, of course, whose unhappiness must derive from purely external, and preferably humanly malevolent circumstances, the smoker who gets lung cancer because of his own habit is less worthy of sympathy than others (at least, unless he can be portrayed as a victim of the tobacco companies); but this, it seems to me, is a very shallow approach to human affairs, by comparison with, say, John Donne's:

> There are too many Examples of men, that have been made
> their own executioners, and that have made hard shrift to bee
> so; some have alwayes had poyson about them, in a hollow
> ring upon their fingers, and some in their Pen that they use to
> write with; some have beat out their braines at the wal of their
> prison, and some have eate of the fire out of their chimneys;
> But I do nothing upon my selfe, and yet am mine owne
> Executioner.

Most (I dare not say all) of us have been our own executioners at some time in our lives; so that not to sympathise with self-executioners is to sympathise with no one.

Be that all as it may, no one could deny that my contemporary who died was the victim of very ill-luck. I do not mean by this that her illness was without cause – it is a basic tenet of a physician's faith that no disease is without cause, even if he does not yet know what it is. I mean rather that the causes that operated on her resulted in unfairness: it was not fair that she should have suffered this fell disease for so long and from so early an age. It was not, however, unjust, because no one willed the unfairness or could, in the present state of knowledge, have prevented it.

The distinction between fairness and justice, and between unfairness and injustice, is a crucial one. Some languages, I believe, do not make the distinction; in French, for example, fair might be translated as *équitable* or *juste*, but neither quite captures the connotations, or even the denotation, of the word fair, at least in this context. One would not say, for example, that it was inequitable for my contemporary to have had this disease, any more than that it was unjust. It is even possible, I suppose – though I do not insist upon the correctness of my wild surmise – that one of the reasons for French statism by comparison with Anglo-saxon individualism is that there is no French equivalent of the word fair in which I have used it: for if you do not have a simple word to describe differences that do not arise from inequity or injustice, you will easily suppose that all such differences can be abolished by political action. (Of course, you can in French express the idea of differences that do not arise from human action, but not so succinctly or monosyllabically as in English.)

Unhappily, people in English-speaking countries seem to me more and more inclined to conflate fairness and justice. Let me give one example (and I mean no disrespect to the author I am about to quote who, an eminent viroloist, seems to me a marvellous man). It comes from near the beginning of his most fascinating memoir. He writes, describing his own passion for both scientific enquiry and social justice that led him to take up clinical virology as a career:

… and ill-health as surely the worst kind of injustice.

But ill-health is not a kind of injustice at all, unless Man by nature is immortal and is only prevented from being so by someone's malevolence. This is absurd: it takes us back to the Azande, the South Sudanese tribe studied by the famous anthropologist, E. E. Evans-Pritchard, who found that they, the Azande, believed that no one died except by the black magic of enemies. (We are fast approaching Azande-levels of intellectual sophistication, for every doctor has experience of the decreasing willingness of people to accept that death is inevitable, and who therefore find the death of a loved one anomalous and *ipso facto* evidence of medical incompetence or worse.)

The human world being so complex, the distinction between unfairness and injustice is not always a straightforward one. Obviously, ill-health can be a manifestation or consequence of injustice, even of gross injustice: for example, there have been many outbreaks of fatal epidemics in prison camps run by vile dictatorships. But ill-health in itself is not pathognomonic of injustice: one cannot say of a person that he is the victim of injustice because he is ill.

Let us take a simple example. It is unfair that I was not born as good-looking as I should have liked. I should like to have been extremely handsome (though I must admit that extreme handsomeness has some drawbacks too, if the extremely handsome people I have known are anything to go by). I did nothing to deserve not being born as I should have wished.

There was nothing unjust in this, and I have no one against whom I might direct my resentment of it. But it is open to me to resent the advantages that accrue to handsome people as against the disadvantages that accrue with a visage such as mine; these are the consequence of human preferences and actions, and hence come, at least potentially, within the realm of justice. Why should handsome people, who have done nothing to deserve their advantage, be received better everywhere?

And so, armed with my resentment, I set about reforming the world: I want to counteract the advantages of the handsome, for example by suggesting that at interviews for jobs, faces should be veiled so as not to create a prejudice against the ill-favoured. In this way, the handsome will not get all the best parts on the world's stage as an undeserved reward for a quirk of nature.

The situation is often much more complex than I have described, of course. Advantages tend to be inherited because parents try to give their children as good a start in life as possible: and this always means a better start than someone else's children who, through not fault of their own, are comparatively disadvantaged. There are parents who either do not know how, or do not care, to give their children the best start in life that they can; and even if they do want to, social circumstances may prevent them from doing so. I doubt there has ever been any society in which the children of those at the bottom of the pile compete absolutely equally with those who are born at the top of the pile.

And so unfairness slides by degrees into injustice: what starts out as being in the nature of things ends up by being remediable, at least in the eyes of reformers.

Sometimes reformers are right; glaring anomalies are susceptible to correction. It is not difficult to find historical examples, nor is it difficult to find examples of necessary reforms in all contemporary societies. Unfortunately, however, reform can easily become a substitute religion, giving meaning to the lives of reformers. As a substitute religion it is not a very satisfactory one, as John Stuart Mill famously intimated in his posthumously-published *Autobiography*:

In this frame of mind it occurred to me to put the question directly to myself: "Suppose that all your objects in life were realized; that all the changes in institutions and opinions which you are looking forward to, could be completely effected at this very instant: would this be a great joy and happiness to you?" And an irrepressible self-consciousness distinctly answered, "No!" At this my heart sank within me: the whole foundation on which my life was constructed fell down. All my happiness was to have been found in the continual pursuit of this end. The end had ceased to charm, and how could there ever again be any interest in the means? I seemed to have nothing left to live for.

Moreover, while justice is desirable, it is not the only thing in human life that is desirable. Injustices should be righted, but not at any cost.

For example, it might be plausibly alleged that the hereditary principle – the principle by which children inherit from their parents – is unjust. Why should you inherit $100,000,000 from your parents, and I nothing, merely by the operation of chance? And, indeed, there are those who would set inheritance taxes at 100 per cent if they could.

Unjust inheritance is not the prerogative only of individuals, however, but of whole generations. What have I done that I should deserve to live forty years longer than people in the middle of the nineteenth century? Should I look with hatred at succeeding generations because, without deserving to, they will live longer, healthier lives than mine, an injustice towards both me and them?

We can now see why equality of opportunity is both a just and a vicious ideal. If anyone were serious about trying to reach it – as against using it rhetorically, whenever handy, to pursue sectional interests – he would quickly destroy everything that makes life worth living, including the love of parents for their children and civilisation itself.

It is important first to distinguish between unfairness and injustice, but it is also necessary to be aware that the righting of injustice has to be weighed against other considerations. It is possible – I think likely – that a totally just society would be a horrible one. One that was fair would be intolerably dull, for it would eliminate difference.

But I cannot think of my deceased contemporary without grief.

40
Strictly for the Birds

F rom the window in my study I can see the bird table in our small garden. Although I am no ornithologist, I can tell a hawk from a handsaw, or rather a thrush from a jackdaw, and the behaviour of the birds amuses me greatly. It sometimes distracts, or perhaps I should say diverts, me from what I should be doing.

Every morning and evening I put out seeds for them. We have reached such a state or refinement of consumerism that even seed mixture sold for the birds is now attractive and even appetising; it looks like muesli for very small people, and is no doubt fortified with all kinds of minerals and vitamins. I think there have probably been times in human history when people were given worse to eat.

Is it really a good thing to feed the birds? Could it make them lazy or incompetent in finding their own provender, so that when through absence or for some other reason we are not able to put out the seeds for them they will be worse off than if we had never assisted them in this way? Are we creating a welfare state for birds, and thereby making them dependent on us?

The table is on a wooden pillar, designed to make it difficult or impossible for rats and squirrels, who are no respecters of human intentions, to climb. On the top there is an arrangement like an open cuckoo clock, except that it is more likely to contain a pigeon than a cuckoo.

The pigeons, it must be confessed, are the major beneficiaries of our largesse: or at least, one of them is. Pigeons are fat and, like so many children in Britain and America these days, they do not know when to stop eating. One of these days one among them – the pigeons, I mean - is going to get stuck in the bird table's shelter, like Pooh in Rabbit's door after Pooh had been true to his inner voice: 'It was just as if somebody inside him were saying, "Now then, Pooh, time for a little something."' For pigeons, as for Pooh, it is always time for a little something.

There is just room for two pigeons on the bird table if they co-operated, and this, paradoxically, gives the small birds their chance. For two pigeons never co-operate and are much more concerned to make sure that the other pigeon does not get any seed than they get some seed themselves. Pigeons are greedy, but greed is not their only vice. Envy is another, and so they chase one another away. It is this that gives the smaller ones their opportunity.

Ideally, of course, every pigeon would like a monopoly of the bird table; but the other bird's deprivation is more important to them than their own satisfaction. Pigeons, then, are human, all too human. They continue their feud in other parts of the garden, chasing their opponents from whatever perch they find them on. While this is happening, the little birds, the sparrows, tits and robins, take advantage.

Alas, their time is short. One of the pigeons usually emerges triumphant, and returns to the table having established his superiority over his fellow Columbidae. He – for I assume it is a male – then occupies the table for up to half an hour, his head bobbing up and down like one of those nodding-donkey oil wells on a plain.

Moreover, when the small birds get their chance, they do not co-operate among themselves. Although there is plenty of room, if not for all then for many, they too do not like to see their diminutive brethren contented or happy, but try to chase them away. I confess that when I saw this, I thought of the Second Balkan War: Bulgaria, Serbia and Greece, having defeated their common enemy Turkey in the first such war, then fell out among themselves and ruined their own economies.

The behaviour of the birds constantly reminds me of human conduct. For example, the other day a carrion crow landed on the table. It was large and black, its heavy beak hardly suited to pecking at small seeds (as inexperienced users of chopsticks cannot pick up a grain of rice), and for some reason the glossiness of its feathers gave me the impression that it was highly pleased with itself, like an unctuous Church of England bishop in the novels of Trollope. It was sitting there complacently when it was suddenly attacked by a furious male blackbird, much smaller than he but full of moral indignation. Beak for beak, of course, the crow was much the more powerful, but he fled before the onslaught. The blackbird then did something rather like a victory roll in the air above the garden.

His victory was the result of his self-belief: a combination of anger and awareness of the justice of his cause. For only a few yards from the table is a wall with some very thick creepers, in which he had his nest. Carrion crows will eat eggs and take young birds; the blackbird was fighting for the life of his young, even if the carrion crow would have disowned any aggressive intention. No, he had come just for the bird-seed; the blackbird was acting upon prejudice, or species-profiling. He was infringing the blackbird's basic ornithological rights. Or, as Mao Tse-Tung might have put it, the carrion crow is only a paper tiger.

It is surprising how quickly we infuse the natural world with our human meanings. We anthropomorphise it almost by second nature. My wife, when she puts out the seed, is often morally disgusted at the conduct of the pigeons, so fat and well-fed already, so greedy, so indifferent to the hunger of others, in a word so thoughtless. Moreover, and this is not the least of it, they seem to defaecate more than any other of the birds, even allowing for their larger size. They are gross, they have no delicacy, no *savoir-vivre*. The other birds are refined, at least when they are not squabbling; they have finesse and delicacy. It is true that the sparrows are not exactly songbirds, having little more than a chirrup in their repertoire; but it is a jolly sound. The blackbird, when he is not angry, is a fine songster. The pigeon's coo is monotonous, utterly lacking in imagination, and he is apt in any case to make a clumsy clattering sound with his wings. His landing is always almost a crash-landing; the phrase 'a bull in a china shop' could be replaced by 'a pigeon in a lilac tree.' Our lovely lilac is often damaged by pigeons attempting to sit in it.

Dislike of pigeons is very common (though, in England, the keeping of pigeons for racing was long a favourite past-time of the working classes). In fact, they are rather beautiful birds, with fine and delicate colouring; our dislike of them causes us to neglect their finer points. It is true that they are not elegantly shaped, and that it is not for nothing that we use the expression 'pigeon-toed' of someone with a certain kind of inelegant gait; but surely everyone has a right to be judged by his best qualities and not his worst?

Is it not strange that, while the word pigeon has so many negative connotations, the word dove has so many positive ones? You'd think, from their reputation, that doves do not defaecate. A dove is a universal symbol of peace; and it was regarded as a sign from heaven, almost, when a dove landed on Fidel Castro's shoulder after one of the first speeches he gave in public as leader of the Cuban Revolution. It would have been completely different if the dove had been a pigeon; and just think of what the outrage would be if, after some ceremony or other supposedly designed to bring about or pray for world peace, the participants released a whole load of pigeons into the air instead of doves! They'd be accused of wilfully procuring damage to public buildings rather than symbolising their desire for a conflict-free world. No one puts wire netting over stone buildings to keep away the doves; they put it there to keep away the pigeons.

And yet there is no clear biological difference or dividing line between pigeons and doves; there is no purely biological criterion by which to say whether a bird is a pigeon or a dove. It is a distinction, then, seemingly without a difference. Doves, it is true, tend to be smaller than pigeons; no one would call the pigeons in my garden doves; yet there are some pigeons that are undoubtedly smaller than some doves, and hence – it follows logically – there are some doves that are larger than some pigeons. In other words, the difference between pigeons and doves is not a biological one, but as our academics would put it, is socially constructed.

The strange thing is that I persist in thinking that I can tell a pigeon from a dove, but this is probably an indication of how prejudiced a person I am.

Be that all as it may, I do not think that my observations of bird-behaviour tells me anything about the human world, though I am so tempted to describe it in human terms. This is not because my observations are unsystematic. I do not think that if I made more systematic observations – for example, if I developed a statistical analysis of aggressive behaviour inter- and intra-species - I would find anything much more about the human world. It is as absurd to study man to find out about sparrow-behaviour as the other way round. One would not conclude from the fact that young men in my town like to get drunk of Friday nights that sparrows, or even robins, like to get drunk on their weekends too.

The list of people who have thought that the examination of the conduct of animals sheds profound light on the human world is a long one. I remember that, in my days as a student, the studies of Konrad Lorenz were all the rage, at least until it was discovered, or at least publicised, that he had been a Nazi. This led to a different, and less favourable or credulous reading of his book on aggression, though of course whether what he said in that book was true or not had nothing to do with his political past.

For quite a long time, when I was a more frequent reviewer of books than I am now, I used to be sent books on ethology and evolutionary psychology for review, and many of them were indeed fascinating. But it seems to me that they shed no light on human life, just as a rigid Laplacean determinism does not help us to live. When I come to a T-junction, it may well be that whether I turn left or right has been already determined by the whole of the previous history of the universe (although it sounds a bit grandiose to put it like this), but the fact is that, when I come to the T-junction, I still have to think about whether I am going to turn left or right. Consultation about the whole of the previous history of the universe will not help me very much, and indeed would turn me into a kind of Buridan's ass.

I do not think it follows from the fact that the komodo dragon hunts in packs that young British youths in the London Borough of Streatham – to take an example purely at random – are predetermined thereby to hunt in packs also. After all, the komodo dragon is a reptile, and Man is not a reptile: except metaphorically-speaking, of course. I think possibly the most foolish extrapolation from animals to man that I have ever read was in the preface that the great ethologist, Robert Trivers, wrote to Richard Dawkins' book, *The Selfish Gene*. In that he extrapolated directly from the bees and the wasps to Man. Oh Man, where is thy sting?

But I shall continue to observe my birds – for the sake of their intrinsic interest, not to find out how to live.

41
Haydn Seek

W hen I am in England I am fortunate enough to live in a pleasant little town which holds the only annual Haydn festival in the country. At lunchtime during the festival I can walk to either of two nearby churches to listen to a chamber concert. The better church, acoustically, was built by Thomas Telford, the great engineer who invented the suspension bridge. His architectural style was cool, classical and rational rather than Gothic or decorated, not at all suited to religious zealotry and more adapted to a tepid deism than to transports of pietism.

It is a perfect place in which to listen to Haydn string quartets, which I love; and this year's quartet was very good. It was formed in 1993 and I was much moved by the evident affection of the players for one another after so many years of ceaseless and unrelenting work in close association together. They were specialists in the Eighteenth and Nineteenth Century repertoire, and surely the inexhaustible depth and beauty of this repertoire was not unconnected to their capacity to survive constant close association with such good feeling. I am not by nature envious, at least by comparison with many other people I have known, but I confess that I envied them.

Haydn is an interesting figure, for he refutes in his own person the romantic notion that a creative person must either be tormented or a swine. (Haydn was tormented, but only by his wife who was a shrew, and that is not the kind of torment that the romantics mean. They mean inner torment.) Haydn is universally acknowledged to have been a delightful man with the most equable temperament; but the virtual inventor of the string quartet, and masterly composer for it, can hardly be denied the title of genius. Mozart deeply and sincerely revered him; there could be no better testimonial than that.

Another undoubted genius of the most attractive character who comes to mind is Chekhov, probably the greatest writer of short-stories who ever lived. No doubt there has been a deal of hagiography in the way that he has been memorialised; but not even the late Christopher Hitchens could have debunked him in even a minimally convincing way. Few men have ever had such an inextinguishable if evasive charm; even Tolstoy, who was not easy to please, least of all by people who were not contented to be his uncritical acolytes and were his equals, loved him.

Personally, I have never been convinced of the supposed link between genius, or even great talent, and bad character. Perhaps I have been unusually fortunate, but the people of real distinction whom I have met, some of the greatest men or women in their fields, have mostly been delightful people into the bargain (though not quite all, but I will not reveal the exceptions). In my experience, it is the moderately talented, those with some talent but enough self-knowledge to wish it were more and worry themselves that it is not, who have a tendency to unpleasantness, for they are disappointed and often bitter that their reach exceeded their grasp.

A friend of mine once told me that a famous Russian writer – Pushkin, I think it was – said that no real genius could be an evil man, that evil was incompatible with genius. I have thought about this question on and off ever since. To decide the question properly, according to the current canons of science, one should have an operational definition of both genius and evil, then select a number of geniuses at random and see whether any of them displayed evil. (The number of geniuses you would have to select for examination of their character would rather depend on the number of evil people you expected in a random sample of ordinary people. The smaller the proportion, the large the number of geniuses necessary. And even then there would remain the black swan problem: because 1000 geniuses had no evil characters among them would not mean that the 1001st genius would not be evil.)

What Pushkin – if indeed it was Pushkin – meant was that there was and is an intrinsic incompatibility between genius and evil. Of course, it is not very difficult to think of geniuses with profound flaws of character: Sir Isaac Newton, for example, was inclined to paranoia, could be cruel, and was not much fun. But no one would call him evil, and quite apart from his brilliance as a scientist he was as capable of great wisdom as of great foolishness.

The term 'evil genius' suggests that people do apply the term 'genius' to the doers of destructive or evil deeds when they reach a certain level of intensity, beyond the capacity of the vast majority of men to commit. The evil genius is not merely wicked, and does not confine his evil deeds to his personal sphere, say an exceptionally cruel murder or two; he is Mephistophelian in his cunning to procure evil on the largest possible scale. The first man that comes to mind of this type is, of course, Hitler, though the last century was exceptionally rich in such men: Lenin, Trotsky, Stalin, Mao, Pol Pot, Abimael Guzmán (the founder of *Sendero Luminoso* in Peru) and several, perhaps many, others. In his biography of Lyndon B Johnson, Robert Caro presents his subject almost in such a light, a worshipper of power for its own sake and without scruple in achieving it.

In more exalted fields of human endeavour, it is difficult to think of geniuses who were truly evil: that is to say, to think of evil composers, painters, writers, scientists and so forth.

There is no doubt, however, that talented people often behave in unpleasant or immoral ways. Whether they do so more than the untalented is open to question; but they are sometimes granted absolution by others on account of their talent. The possession of this talent is seen as a kind of burden to be borne, for which the freedom to behave badly is a compensation or consolation, a hydraulic necessity to release the build-up of creative tension, and without which the talent would be unable to express itself fully. In order to benefit from the products of talent, therefore, the world just has to exonerate a certain amount of bad behaviour from the talented.

The first thing to note about this is that there is probably a dialectical relationship between the bad behaviour of the talented and the exoneration of it by admirers of talent. The talented, knowing that they are granted more leeway than others, take advantage of it and behave worse as a result than they might otherwise have done. This can be in large things as in small; and indeed they come to see their own talent as justification or exculpation in advance for their intended misconduct.

But does misconduct affect the way in which we react to the productions of the talented? I recently heard a story from an eye-witness about the behaviour of a late journalist, undoubtedly of very great talent,

that lowered him in my estimation far more than any intellectual dis-
agreements I might have had with him had ever done. He was, I learnt
from this eye-witness, rude and condescending to, and dismissive of, his
social inferiors, especially those who performed services for him. Of all
human qualities, this seems to me to be one of the most disagreeable,
and to reflect worst on a person's character. As the eye-witness to this
behaviour had no axe to grind, and might rather have been expected to
evince admiration for this journalist, I believed what she told me.

But what difference did this knowledge of his character make to
my judgement of his work? His wit was just as witty, his facts as accurate
or inaccurate, his deductions from them just as valid or invalid, as they
had been before. And yet it coloured everything for the man had been a
theoretical egalitarian, outraged, at least in print, by the injustices, ineq-
uities and inequalities of the world.

Hazlitt has an essay on the difference between cant and hypocrisy.
Cant, he says, is expressing a feeling more strongly than one truly feels;
while hypocrisy is espousing principles in which one does not believe:

> We often see that a person condemns in another the very
> thing he is guilty of himself. Is this hypocrisy? It may, or it
> may not. If he really feels none of the disgust and abhorrence
> he expresses, this is quackery and impudence. But if he re-
> ally expresses what he feels (and he easily may, for it is an
> abstract idea he contemplates in the case of another, and the
> immediate temptation to which he yields in his own, so that
> he probably is not even conscious of the identity or connex-
> ion between the two), then this is not hypocrisy, but want of
> strength and keeping in the moral sense.

In Hazlitt's view 'the greatest offence against virtue is to speak ill of
it;' and 'to recommend certain things is worse than to practise them.' On
this view, then, an egalitarian who behaves badly towards subordinates
is better than a believer in inequality who treats subordinates as equals
(if egalitarianism is considered a virtue in the first place).

I do not think this can be quite right, for it makes the expression of
the right ideas the major part of virtue; in other words, virtue becomes
a doctrine rather than a discipline, and we can simply write off the bad
behaviour, no matter how great, of those who espouse virtuous views
as mere weakness of will, if the kind from which all of us suffer. A man
who said he believed in politeness but was never polite would be sus-

pected of lying; only someone who said he believed in politeness and was sometimes impolite but often polite would be held to suffer from weakness of will.

If my journalist's disdain for subordinates were habitual rather than occasional – as my eye-witness, who met him on several occasions, suggested that it was – then his professions of egalitarianism were insincere. But even this would not prove that egalitarianism were wrong, only that he did not truly believe in it. The work is distinct from the man.

It is this distinction that assists the talented in their career of bad behaviour (if they exhibit it). For it is likely that they believe that their work is of great importance for humanity, greater importance at any rate than that of many men; so that their reputation finally relies on their work rather than on their conduct. That being the case, they have more leeway than others to behave badly.

Moreover, the difference between the significance of the work and conduct is likely to increase with time, at least if the work survives the death of its author. If it were to be shown conclusively from impeccable sources that Shakespeare had been a villain all his life, it would hardly affect our estimation of his work at all. A man can be a sublime artist but an unattractive figure, and in the long run it is the former that counts.

I was faced with this problem once when I was writing about Arthur Koestler, a man whose work and intellectual capacity and vigour I greatly admire. He might not have been right about everything, but he was dull about nothing. Yet it was revealed, and widely accepted, that in his private life he had behaved reprehensibly, even criminally, towards women. When I read him now, the word 'Rapist' echoes through my mind. What weight was I to put on his behaviour in the assessment of his work?

That is why the figure of Haydn comes as such a relief to me. A very great, if not perhaps a supreme, artist, he proves that it is possible to do brilliant work and yet be a good man. He would never have believed that his quartets excused, exonerated or mitigated misconduct towards others.

42
A Doctor Writes

A mong my favourite books – I mean the books that I actually
own – is a first edition of Dr Johnson's *Lives of the English Poets*,
published in 1781 in four volumes. I bought the *Lives* in a bookshop in
Dublin – in Blackrock, to be exact – for what seemed to me a bargain
price, though whether it really was a bargain I shall not know until I try
to sell them, which I never shall.

Among the fascinations of these volumes are the annotations
made by the first owner of them, in a clear and elegant, although slightly
spidery hand. Was eighteenth century ink, I wonder, brown when first
put to paper, or did it brown with age? On inspection, the annotations
(with a few exceptions, those being pedantic corrections of the slight errors inevitable in a work 2000 pages long) were actually emendations of
the text. It was Mrs Valerie Myers, wife of the literary biographer, Jeffrey
Myers, and a literary scholar in her own right, who suggested to me that
the emendations might be taken from a subsequent edition of the *Lives*:
and so it proved when I compared the first edition with a subsequent
edition contained in Johnson's *Complete Works*.

This discovery naturally excited me. Could it be that I had alighted
on Johnson's own copy, which he had emended himself for the second
edition? I doubted it; he was very short-sighted, and was no respecter of
books as physical objects. These were in very good condition; Johnson, I
suspect, would have destroyed them in writing the emendations.

But could he have dictated the changes to an amanuensis? This thought set my mind fantasising. Who could the amanuensis have been? A name came at once to mind: Edmund Malone, the Irish Shakespearean scholar who was a close friend of Johnson's. That was why this copy was in Ireland.

Alas, the explanation was quite otherwise, and much more banal. I found a letter in the fourth volume that explained it all. It was from a scholar at Trinity College to a former owner of the books, explaining that when the second edition of the *Lives*, with emendations, was brought out, purchasers of the first edition were invited to booksellers who would give, or sell, them a pamphlet containing all of Dr Johnson's emendations. Apparently only ten copies of this pamphlet were ever taken up; but the original owner of what was now my copy had conscientiously copied the emendations into the four volumes.

This story brought to mind one told me by an historian at Oxford. In the 1930s, the Bodleian bought a copy of the *Great Soviet Encyclopaedia*. With changing political circumstances, entries about various prominent figures such as Bukharin and Yagoda became inoperative, to use a later political euphemism; and purchasers of the encyclopaedia were circulated with a letter from the publisher asking them to tear out the original page and insert the enclosed, 'corrected' page (the entry on Yagoda being replaced by one on the yak, or some such). The real point of this story was not to illustrate Stalinist dishonesty, which is sufficiently well understood, but to illustrate the moral weakness or ideological leanings of some of the Bodleian staff: for they complied with the instructions given, and duly replaced the pages.

But let me return to the *Lives*: this is a book that I love for its content as well as its form. Johnson peppers his biographical sketches with moral observations of great interest. One of my favourites is in his life of Swift, a man he did not altogether admire. Remarking on Swift's eccentricity, Johnson says:

> Singularity, as it implies a contempt for the general practice, is a kind of defiance which justly provoked the hostility of ridicule; he therefore who indulges in peculiar habits is worse than others, if he be not better.

If he be not better: it is in this clause that Johnson shows the superiority, seriousness and honesty of his mind, for a lesser writer, a mere reactionary hack, would not have added it. Johnson is therefore not saying that all change is to be reprehended, that custom and convention must be our guide through thick and thin; but neither is novelty to be welcomed or applauded for its own sake.

Referring in his life of Thomas Gray to his subject's travels, Johnson says:

> It is by studying at home that we must obtain the ability of travelling with intelligence and improvement.

The uninstructed traveller gawps; the instructed observes. One is reminded of Pasteur's later remark that chance favours the mind prepared.

Of some of the stanzas of Gray's famous Elegy, Johnson says:

> I have never seen the notions [in them] in any other place; yet he that reads them here, persuades himself that he has always felt them.

This is precisely the great quality of Johnson's aperçus: that he who reads them persuades himself that he has always perceived their truth, though he never has.

Johnson is not very complimentary about Gray's poetry on the whole, and indeed he makes sport of some of its infelicities. But he allows the Elegy a full measure of greatness; for, says Johnson:

> ... by the common sense of readers uncorrupted by with literary prejudices, after all the refinements of subtility and the dogmatism of learning, must be finally decided all claims to poetical honours.

Take that, all you literary theorists!

Curiously, some of the sentiments expressed in the Elegy, a few of whose verses run through my mind in every cemetery (and I love cemeteries, I can resist them no more than bookshops), are at variance with the slight faults of Gray's character that Johnson enumerates in his biographical sketch. Compare for example the following eight lines, whose generous sentiments towards the humble people buried in the churchyard are incomparably expressed:

> Let not ambition mock their useful toil,

Their homely joys, and destiny obscure;
Nor grandeur hear with a disdainful smile
The short and simple annals of the poor.

The boast of heraldry, the pomp of power,
And all that beauty, all that wealth e'er gave
Awaits alike th'inevitable hour –
The paths of glory lead but to the grave...

Compare them, I say, with what Johnson says of Gray – quoting Gray's intimate friend, Mason:

There is no character without some speck, some imperfection; and I think the greatest defect in his was... a visible fastidiousness, or contempt and disdain of his inferiors in science.

But which is more significant? The verses I have quoted, or the speck upon his character? Surely, nearly a quarter of a millennium after his death, the former.

One of the lives that I love most, and can re-read with great pleasure and instruction, partly for medical reasons, is that of Richard Savage (that Johnson originally published separately many years earlier). Savage was a minor poet and playwright whom Johnson befriended in his youth; and I doubt that a better portrait of a charming psychopath has ever been written.

Savage belonged to that small and select group of writers who were once condemned to death and reprieved: I can think, offhand, of Dostoyevsky, Koestler and the greatest South African writer (not much known or appreciated outside South Africa, however), Herman Charles Bosman. Savage was involved in a sordid quarrel with a man in a tavern and ran him through with his sword. Initially sentenced to death, he was reprieved by the intercession of the Countess of Hertford, but taverns remained his natural habitat for the rest of his life. Johnson captures precisely the failure of psychopaths to learn from experience, all the more powerful because the notion of the psychopath was not yet known:

By imputing none of his miseries to himself, he continued to act upon the same principles, and follow the same path; was never made wiser by his sufferings, nor preserved by one misfortune from falling into another. He proceeded through-out his life to tread the same steps on the same circle; always applauding his past conduct, or at least forgetting it, to amuse himself with phantoms of happiness, which were dancing be-fore him; and willingly turned his eyes from the light of rea-son, when it would have discovered the illusion, and shewn him, what he never wished to see, his real state.

He was a consummate sponger on others, appearing 'to think him-self born to be supported by others, and dispensed from all necessity of providing for himself.' He lived for the moment; 'an irregular and dissipated manner of life had made him the slave of every passion that happened to be excited by the presence of its object, and that slavery to his passions reciprocally produced a life irregular and dissipated.' He was completely unreliable; 'he was not the master of his own motions, nor could promise any thing for the next day.' His 'friendship was of little value; for though he was zealous in the support or vindication of those he loved, yet it was always dangerous to trust him, because he con-sidered himself as discharged by the first quarrel from all ties of honour or gratitude; and would betray those secrets which, in the warmth of confidence, had been imparted to him.'

He was very amusing in conversation and his charm made him friends easily; he was invited into people's houses; but 'another part of his misconduct was the practice of prolonging his visits to unseasonable hours, and disconcerting all the families into which he was admitted.' He never regarded a loan as something to be repaid, and indeed thought it was a clear sign of bad character if anyone asked him for a repayment; not surprisingly, perhaps, he died in a debtors' prison, probably of gaol fever, that is to say typhus.

But if he was a psychopath, he showed traits which are by no means uncommon in the literary world. He was willing to satirise in print those whom he had recently praised in person; and I remember once being the victim of this kind of behaviour at the hands of an eminent novelist, to whom I was introduced at a party. He had read some of my articles, and it would hardly be an exaggeration to say that he fawned on me and, being naive in the ways of the literary world, I was much flattered. Two weeks later, perhaps at a loss for something to say in his weekly newspa-per column, he called upon the authorities, to a readership of millions, to drum me out of the medical profession (my livelihood). Ever since, I have listened with reserve to praise uttered by writers.

But honesty compels me to admit that I, in common I suspect with many scribblers, exhibit one of the traits of Richard Savage as described by Dr Johnson.

> But though he paid due deference to the suffrages of mankind when they were given in his favour, he did not suffer his esteem of himself to depend upon others, nor found anything sacred in the voice of the people when they were inclined to censure him; he then readily shewed the folly of expecting the public to judge right, and was somewhat disposed to exclude all those from the character of men of judgement who did not applaud him.

However:

> He was at other times more favourable to mankind than to think them blind to the beauties of his works, and imputed the slowness of their sale to other causes; either they were published at a time when the town was empty, or when the attention of the publick was engrossed in some struggle in the parliament, or some other general concern; or they were by neglect of the publisher not diligently dispersed, or by his avarice not advertised with sufficient frequency. Address, or industry, or liberality, was always wanting; and the blame was laid rather on any person than the author.

Ah yes, I've been there and thought all that. It's uncanny, really. Savage was always able to live at peace with himself, says Dr Johnson, explaining it with that largeness of mind and generosity of spirit that was characteristically his:

> By arts like these, arts which every man practises in some degree, and to which too much of the little tranquillity of life is to be ascribed...

Man does not live by truth alone.

CPSIA information can be obtained at www.ICGtesting.com
Printed in the USA
LVOW132205250812

295938LV00004B/3/P